PRACTICAL MANUAL OF CHEMICAL PLANT EQUIPMENT

PRACTICAL MANUAL *of* CHEMICAL PLANT EQUIPMENT

by
ROBERT G. SCHMIDT, P. E.
Columbia Nitrogen Corporation
Augusta, Georgia

CHEMICAL PUBLISHING COMPANY, INC.
New York
1967

Practical Manual of Chemical Plant Equipment

ISBN: 978-0-8206-0086-4

Chemical Publishing Company:
www.chemical-publishing.com
www.chemicalpublishing.net

First edition:

© Chemical Publishing Company, Inc. – New York, 1967

Second Impression:

© Chemical Publishing Company, Inc. - 2011

Printed in the United States of America

Preface

This complete manual of commonly used chemical plant equipment has been written for the chemist, engineer, operator, or mechanic entering the chemical industry. The manual is designed to explain the principles of operation; advantages and disadvantages of the different types of equipment; terminology used in the industry; and some of the important basic operating theories and practices.

When first viewed, most chemical plants are a confusing maze of pipe, valves, pumps, columns, tanks, and other equipment that appear to be impossible to understand. The language commonly used by the "old hands" is also very confusing to the new man. The purpose of this manual is to provide knowledge of the commonly used equipment and terminology so that the new man can be relieved of much of his initial mental confusion. He can thus devote his efforts to the study of the special equipment used and grasp the overall idea or sequence of operating steps so necessary to an understanding of the particular plant and the job to be accomplished.

March 1967

ROBERT G. SCHMIDT

Contents

1. SOLIDS—STORAGE AND HANDLING

GENERAL

It is still quite common to find solids being received, stored, and handled in bags or fiber drums. For large quantities of solids, however, the hopper car, some type of conveyor, and bins or silos are more economical. In many places in the process, the solids might be present with sufficient liquid as a slurry that can be pumped. (Generally less than 25 percent solids by weight can be pumped successfully.) This section of the manual will deal with solids where there is insufficient liquid present to use a pump.

CONVEYORS

Air Conveyor

The air conveyor has become extremely popular and it can be used on almost all but wet sticky materials, although some breakage will occur in the material being conveyed.

There are four general types:

A *suction* or negative system sucks the material up, much as does a vacuum cleaner; this system is used especially where material must be picked up from a number of points and deposited at one or more points. Only one operation can be performed at a time.

The *pressure* or positive system moves the material by direct pressure created by a blower located ahead of the intake of material. This system is used where there is one intake place and several delivery points.

The third system is a combination of the two, shown in Figure 1:1. Note that the blower can suck the material from the hopper car to the storage bin and can also blow the material from the bottom discharge of the silo to another point.

The fourth system, called *fluidizing*, is accomplished by mixing

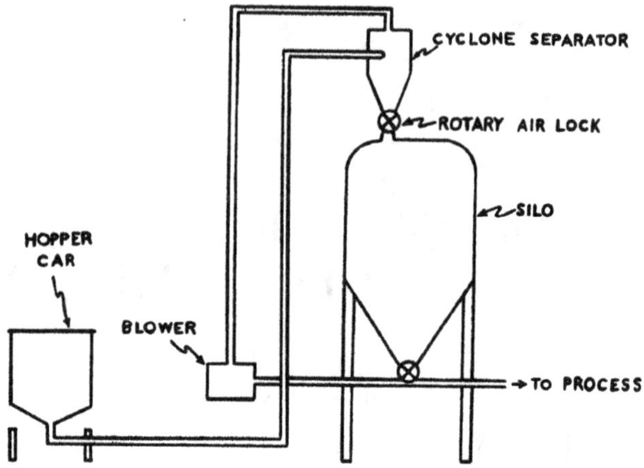

Fig. 1:1. Combination suction and pressure air conveying system.

each particle in a thin envelope of air. This system moves a great deal of material with a minimum of air and minimum of degradation or breakage. For example, 20 to 30 standard cubic feet per minute (SCFM) of air moves about 1,000 lb. an hour through a one-inch pipe. Actually, the quantity of air would depend upon the length, number of bends, nature of material being conveyed, and height to which it is to be conveyed.

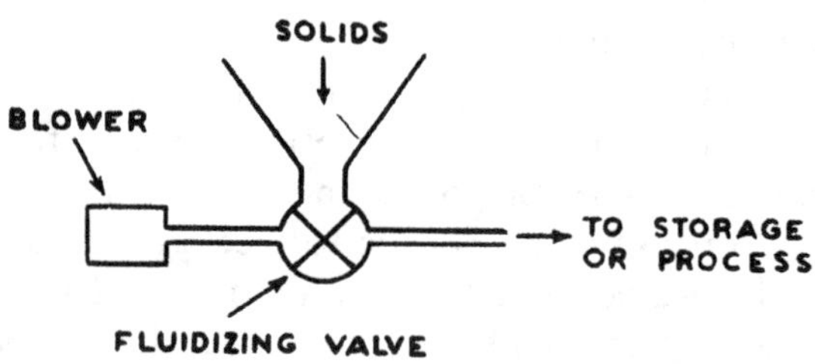

Fig. 1:2. Fluidizing type air-conveying system.

Figure 1:2 is a simplified diagram of the fluidizing portion of a fluidizing system. The blower supplies air to a rotary air-lock valve. The rotating blades can regulate the rate of solids being fed into the system, mix the solids and air, and take them over to the discharge pipe. A fluidizing valve is needed at each point of pickup of solids.

As with the other air conveyors, the pipe discharges into a cyclone separator and through a rotary air-lock as shown in Figure 1:3. The cyclone separator is a vessel or piece of pipe of large diameter and the conveying pipe discharges into the cyclone in a tangential way to give the air and solids a circular motion. The air then rises at a slow velocity because of the large diameter of the cyclone, thus allowing the solids to drop down into the rotary air-lock. The rotary air-lock lets the solids drop into the silo or bin and keeps the air from entering.

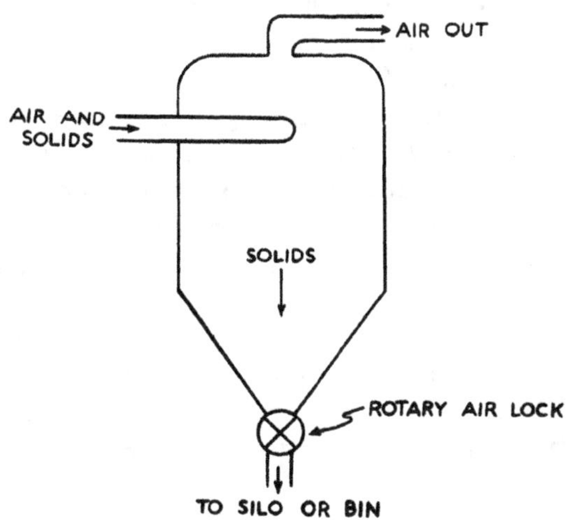

Fig. 1:3. Cyclone separator.

As a cyclone is not 100 percent efficient, the air will contain about one percent of the solids fed into it originally. This air is frequently taken to a second cyclone, bag filter, or other type of collector or scrubber to recover these solids.

Belt Conveyor

One of the most economical types of conveyor is the belt conveyor; however, their use is limited to reasonably level straight runs. The angle at which a belt conveyor may be run depends upon the material to be conveyed.

Vibrating Conveyor

In the vibrating conveyor the solids lie in a trough, which is vibrated in such a way as to make the particles move along in short jumps. It is not good for wet sticky materials and is generally used only for short, straight, horizontal, or downhill runs. A spiral, elevating, vibrating conveyor is available but it is rather expensive.

Zipper Conveyor

A zipper conveyor moves with the solids as a belt conveyor does, but it has the advantage of being able to elevate material and change direction. The maintenance costs are relatively high compared with other types and fine rubber particles gradually created by wear could cause contamination of certain materials.

En Masse Conveyor

The *en masse* conveyor consists of a chain holding a succession, of solid paddles, or vanes, often called *flights*, that move with the material being conveyed. It is so called because it is designed to run completely filled, and the chain, the flight of paddles, and the solids move together. In a newer type, called a *modified flight*, the vanes or paddles are not solid, but have a large slot in each. This type uses less power and is less noisy than the other form, but it can carry only solids that can interlock with one another—nothing that might fall through the openings.

The major advantages of *en masse* conveyors, with both solid and modified flights, are that they can operate under positive pressure, they can elevate solids, and they can change direction. They are, however, generally a little more expensive than other types of conveyor.

Screw Conveyor

This well-known screw type of conveyor is satisfactory for most materials but it is generally noisy and limited to straight short runs.

Bucket Elevators

The bucket elevator is used where elevation of the solids is the only requirement. There are three major types:

1) A *centrifugal discharge elevator*, which must operate at a relatively high speed.
2) The *continuous bucket elevator* in which, at the discharge point, each bucket drops its content on the sloping bottom of the bucket below, deflecting the solids into a discharge chute.
3) The *positive discharge elevator* in which the chain passes over the upper sprocket and then passes two snub sprockets so that the bucket is turned through more than 180 degrees.

A major problem of bucket elevators is the slow rate of movement of solids and the relatively large height needed at the loading and discharge ends of the conveyor.

Feeders

Where it is necessary to supply solids to a process at a controlled rate, a *volumetric* feeder (which delivers a certain volume of solids), or a *gravimetric* feeder (which delivers solids by weight) is used.

Most of the conveyors previously listed can serve for volumetric feeding. A rotary valve, sometimes called a *star feeder* is an economical model to use and is shown in Figure 1 : 4.

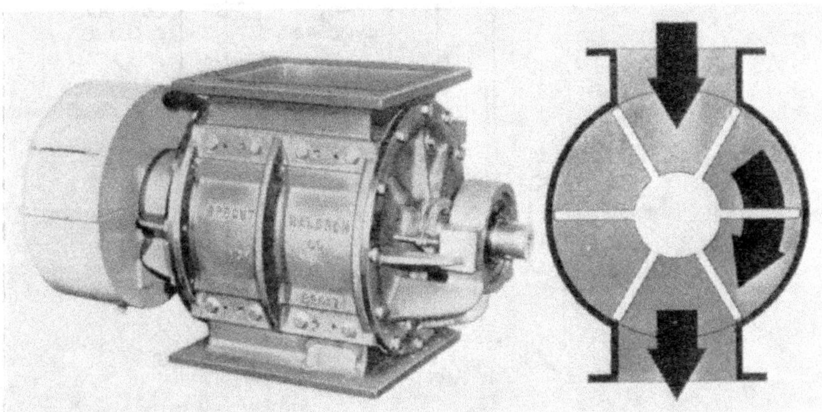

Fig. 1:4. Typical star feeder. (*Sprout Waldron*).

Gravimetric feeders are used where the rate of flow by weight must be accurately controlled. Generally, these consist of a belt moving over a weighing scale and the speed of the belt, or the amount of solids fed to the belt, is adjusted to obtain the desired weight of solids delivered.

DRYING

The most economical method of removing liquids from solids is by mechanical means, such as the use of filters and centrifuges. These mechanical methods generally leave from 2% to 50% liquid with the solids, which must be removed by some type of dryer.

Flash Dryer

The flash dryer is one of the most popular dryers where a powdered product is desired. As it employs air-conveying and short residence time it is not too satisfactory for large crystals or where breakage is to be avoided. Some of the variations are the *air-stream flash dryer*, *the cage mill,* and the *recycle.* Figure 1 : 5 shows an air-stream type in which a relatively dry material is to be fed to the flash dryer and breakage of the product is to be minimized.

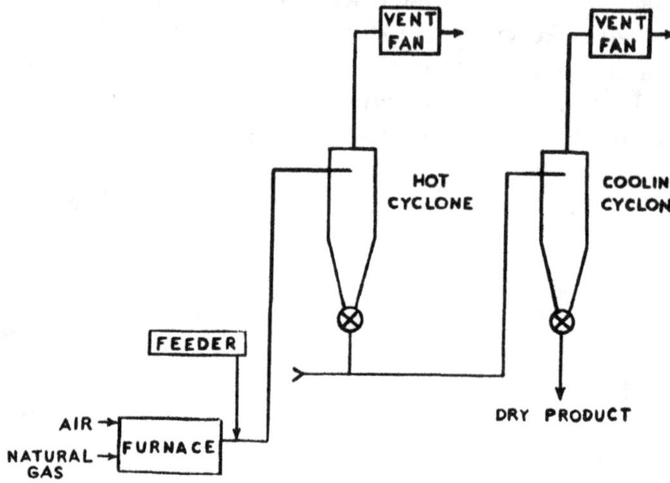

Fig. 1:5. Air-stream type of flash dryer.

Air and natural gas are burned in a furnace to provide hot air around 1300 degrees F. Air could also be blown over steam coils but the higher the temperature of the air, the more efficient the system. A feeder, such as a screw-feeder, supplies the wet product at a uniform rate. The product is carried by the stream of warm air up to the hot cyclone and may be at a temperature around 300 degrees F (more, or less, as desired). If the product is then to be cooled, it may be passed into a cyclone system shown in Figure 1 : 6, where cool air is mixed with the hot product.

If the final product is to be very fine or in powdered form, a cage mill can be added. The cage mill consists of rods fastened to a rotating disc which break up the particles of product. Other types of mills are also used—for example, an *impulse mill.*

If the product is wet enough to be like paste, a recycle system can be added in which a percentage of the dry product from the hot cyclone is mixed in a paddle-blade mixer with the incoming wet material. A complete system using a cage mill, a recycle, and a cooling cyclone is shown in Figure 1 : 6.

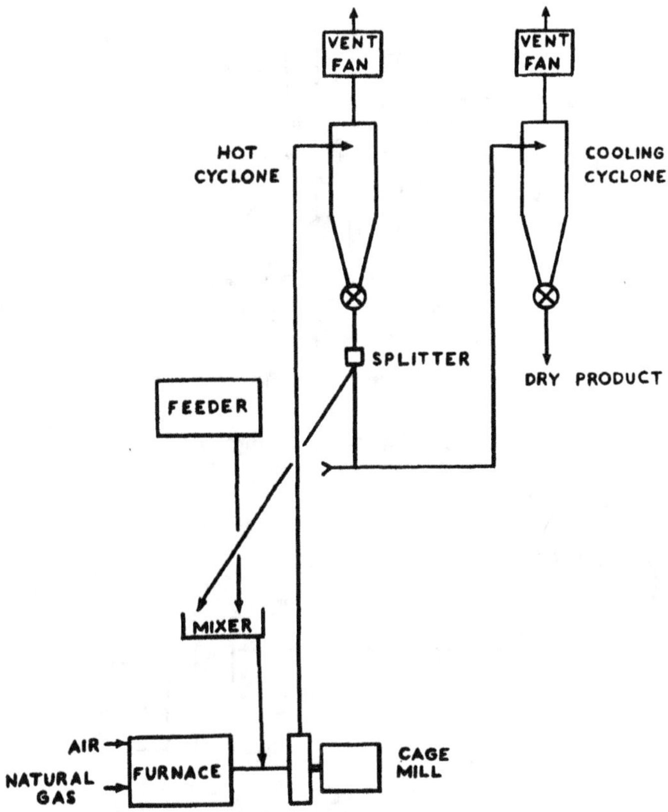

Fig. 1:6. Flash-drying system for producing dry powder from a paste feed.

Fluid Bed Dryer

Where breakage of a product must be minimized and longer drying time is needed than is provided by a flash dryer, the fluid-bed dryer can be used. It is similar to the flash dryer, and explained in Figure 1 : 7.

Rotary Dryer

One of the oldest and most widely used dryers is the rotary dryer. A simplified version of a counter-current dryer is shown in Figure 1 : 8.

A feeder supplies the wet product to a rotary joint at the head of the rotating drying cylinder. Vanes in the cylinder lift and turn the product over and gradually move it to the discharge end. A furnace

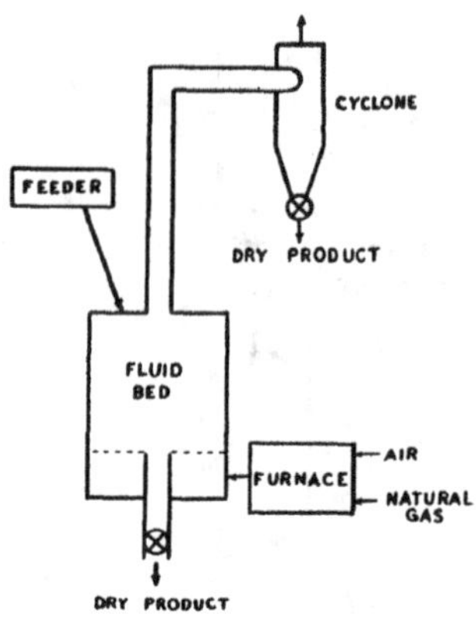

Fig. 1:7. Fluid-bed drying system.

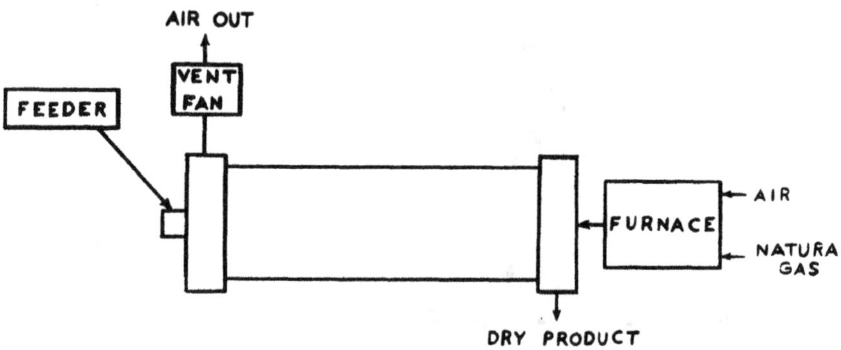

Fig. 1:8. Counter-current rotary dryer system.

forces warm air counter-current to the product. Many variations of this unit are available.

Tray Dryer

Where a small quantity of product must be dried it is often spread on trays, which in turn are frequently placed on portable racks. The trays may be placed in enclosures where warm air is circulated or from which air is pulled to create a vacuum and dry the material. For large volumes of product, drying on trays is not economical unless it is mechanized.

Units are available where the trays rotate about a center shaft and the product is dropped from the upper tray to the next below by means of wiper arms and slots or by a tilting mechanism on the tray. A cross section is shown in Figure 1 : 9.

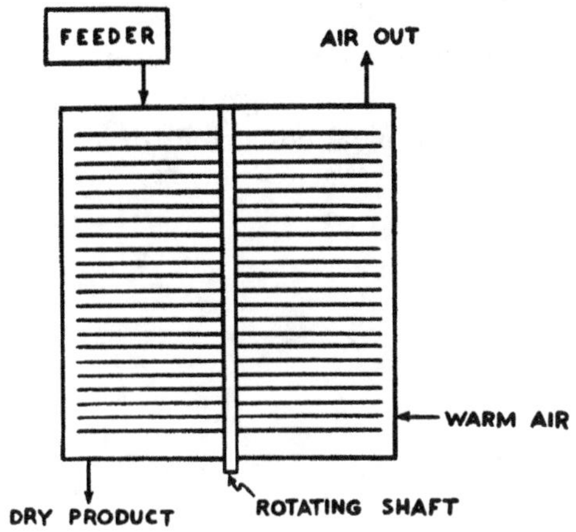

Fig. 1:9. Rotating tray dryer.

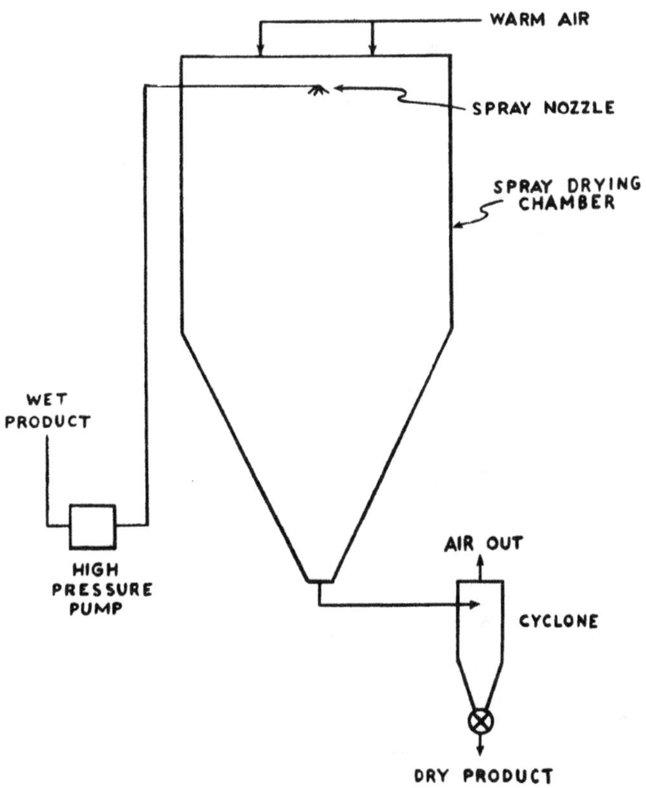

Fig. 1:10. Co-current spray drying system.

Spray Dryer

For the undried product in liquid form or very dilute, or where drying conditions must be very gentle, a spray dryer may be used. Figure 1 : 10 shows a typical co-current dryer.

2. LIQUIDS—STORAGE AND HANDLING

TANKS

Quantities of liquid up to 1,000 gallons are generally stored in vertical tanks mounted on legs. Quantities from 1,000 to 10,000 gallons are generally stored in horizontal tanks mounted on concrete supports. Larger quantities are most often stored in vertical tanks resting on a concrete foundation. The reason for elevating the tank is to obtain a head or pressure at the suction of the pump, known as the *net positive suction head* (*NPSH*) and explained under Pumps in this section.

Many storage tanks are quite simple. A water-storage tank, for example, might be satisfactory when provided with a supply line, a line near the bottom to a water pump, and an overflow line near the top. The overflow line will let air out as the tank is filled, let air in as the tank is emptied, and let water out if the tank is over-filled. Some chemicals are not as easily stored and call for more complicated storage systems.

It is observable that water and gasoline evaporate, that gasoline evaporates more quickly. This is because the vapor pressure of gasoline is higher. The vapor pressure increases as the temperature rises. A tank that is open to the atmosphere has both air and vapor above the liquid in it. As the temperature rises, the amount of vapor increases and both the vapor and the air expand. Some air and vapor are then pushed out of the tank into the atmosphere, or *vented.* As the temperature decreases, vapors condense and both air and vapor contract, drawing in air.

This is called "breathing" in a storage tank, and it goes on as the temperature and atmospheric pressure change during the day and night. The vapors that leave the tank are lost, and if the vapor pressure of a chemical is high, these losses can amount to a considerable quantity of material. If the tank were not open to the atmosphere, it would require heavy walls so that it would not burst when pressure increased, or collapse if a vacuum formed. As heavy-walled tanks are expensive, it is sometimes better to take small vent losses. In addition to the loss of chemical by the "breathing" of the tank, there may be other reasons against an open tank. The escaping vapors of some chemicals may be toxic, or flammable, or have bad odors; other chemicals must be protected from contact with the air or moisture, which might harm their quality.

Holding a tank under one or two pounds of pressure might reduce the loss of chemical to a reasonable amount. If air can be admitted and vapors allowed to go out into the atmosphere, a light-walled tank can be used. It should be protected by a relief valve and a vacuum-breaker.

The tank could be fitted with a special balloon device inflated by the pressure in the tank. If the pressure in the tank becomes too high the balloon pushes open a trap door and lets air and vapor out. If the tank starts to evacuate the balloon collapses and lets air into the tank. When the pressure in the tank is between atmospheric pressure and one or two *psig* (pounds per square inch gauge) the balloon is sufficiently inflated to block the intake of air but not sufficient to open the trap door.

When a tank must be held at a few pounds pressure but air must be prevented from entering the tank, a low-pressure oil-bubbler system can be used. A typical installation is shown in Figure 2 : 1.

When the tank is being filled, the gas—nitrogen—is forced out into the vent header. The same vent header can serve a number of tanks. Gas from the vent header can bubble up through the oil and escape into the air if the pressure in the tanks and vent header is greater than the pressure created by the level of oil. If the tank is being emptied gas flows from the vent header into the tank. If a large quantity of gas is required and the pressure in the vent header drops, the pressure controller (PC) will detect this and open the control valve, allowing additional nitrogen to enter the system. Note that the lines entering the vent header rise above the header and then drop into it so that any liquid that might collect will not run down into another tank. If the gases or vapors given off should be prevented from entering another tank, separate vent headers are required.

Where the tank must be held under high pressure it can be installed as shown in Figure 2 : 2.

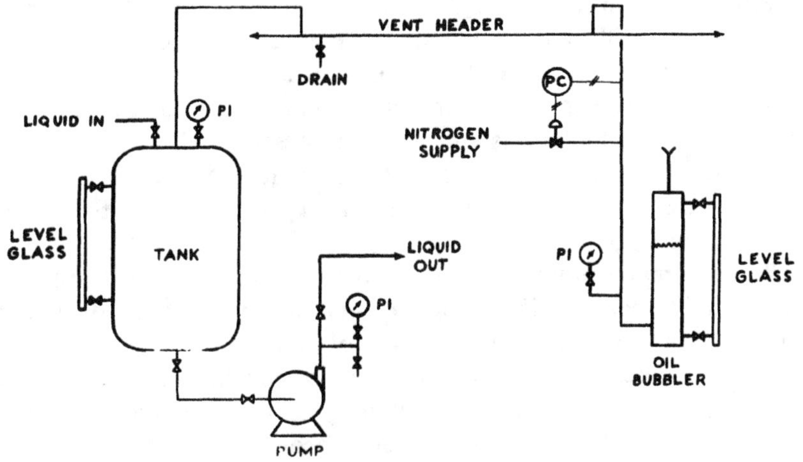

Fig. 2:1. Typical oil bubbler system.

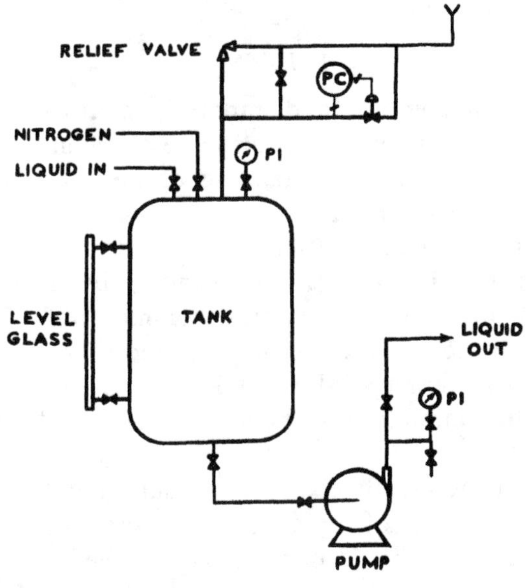

Fig. 2:2. High-pressure vent system.

In the system shown in Figure 2 : 2 air can be eliminated, if necessary, by sweeping nitrogen through the system. As the tank is filled the pressure will increase. If the pressure becomes too high or gas is generated in the tank as it might be if the tank were a reactor, the pressure controller will detect this and let some gas escape. This tank is provided with a relief valve that is set a little higher than the normal operating pressure expected in the tank and is not used except to protect the tank. If the tank is emptied, the pressure will go down. Gas may be added manually or automatically if it is desired to keep the pressure above a certain figure. (See *Pressure Controller* in the section on instruments.)

There are many other considerations in storage of liquids, especially in regard to safety. If the material is flammable—and depending upon the quantity held—the location of the tank, the distance between tanks, dikes, and other items must be considered. If the vapors are toxic or have a bad odor they require treatment before entering the atmosphere. A study of the physical properties of the chemicals used in the plant is important for anyone working in a plant, and a good safety record requires that all operating personnel understand and appreciate the hazards of the chemicals with which they may be working.

PUMPS

One of the most common and important pieces of equipment in a chemical plant is the pump. It is also one of the greatest sources of difficulty and a thorough understanding of the types available and operating problems is essential.

Net Positive Suction Head (NPSH)

It is obvious that liquid must enter a pump before the pump can work on it, but this is one of the most common difficulties with the pump. The ability to get liquid into the pump is generally called the *net positive suction head* (NPSH). It is the pressure in *psia* at the suction side of the pump, minus the vapor pressure of the liquid that is being pumped.

There are two sources that supply pressure and two that reduce pressure. The first source of supply is the pressure of the liquid, which varies with the depth of liquid in the tank that feeds the pump. The second is the pressure of the gas above the liquid. If the tank is open to the atmosphere, the additional pressure will be 14.7 *psi*.

The first loss of pressure is from the flow of liquid through the pipe between the tank and the pump. If the pipe is large in diameter, fairly straight, and has the minimum number of valves, the drop in pressure should be small. The second loss is from the vapor pressure of the liquid that is being pumped. In this manual, the problem of vapor pressure is discussed under *Distillation* in Section 7.

If vapors are allowed to form in any quantity in a pump, the pump will be filled with vapor and pumping vapor instead of liquid. The efficiency of most pumps is greatly reduced under this condition and the effect of it on various types of pumps will be covered as the individual types of pumps are discussed. We must therefore always keep in mind the importance of NPSH. The inability of a pump to develop *head*, or discharge pressure, may derive from any of the following:

Lack of pressure in the tank
A low level of liquid in the tank
A valve partially closed
Solids in the suction line creating a big pressure drop
Air or gas trapped in the pump
A high temperature of the liquid in the pump, causing high vapor pressure

Pumps require a certain NPSH to perform as promised by the manufacturer. This NPSH requirement is generally stated in *psia*, or "*feet of head*." The latter refers to the pressure of the *head*, that is, the height in feet of a column of water. To conver "*feet of head*" of water to pounds per square inch, divide feet of head by 2.3. The figure for feet of head is also generally absolute pressure; therefore, when dividing by 2.3 the result will be in *psia*. The discharge pressure is also generally given in "*feet of head*", or *psia*.

Centrifugal Pumps

Centrifugal pumps are the most commonly used because of lower cost, uniform flow, low maintenance cost, and ability to handle some undissolved solids. If liquids are to be pumped to greater heights or to pressures higher than 100 *psig*, other types of pumps become more economical.

The centrifugal pump consists of a rotating impeller inside a case. The liquid enters the center, or eye, of the impeller and the blades of the impeller throw the liquid out, giving the liquid a centrifugal force as shown in Figure 2 : 3.

There are three types of impeller in common use: *open*, *semi-open*, and *closed* (Fig. 2 : 4). The closed impeller is the most efficient, but where some solids may be expected the semi-open or open impeller is less likely to become plugged. It is becoming customary to simply

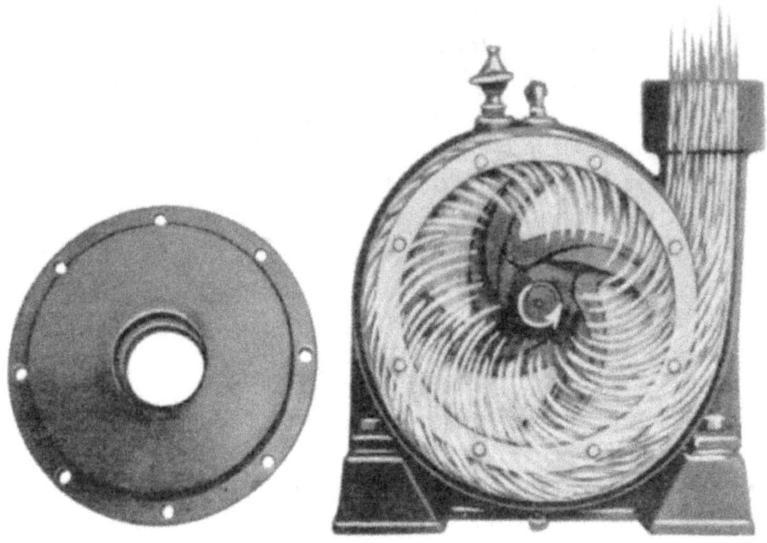

Fig. 2:3. Centrifugal pump (*Goulds Pumps*).

say *closed* impeller when there is a front and back plate or shroud to hold the blades, and to refer to all others as *open*.

A rather typical piping arrangement for a centrifugal pump is

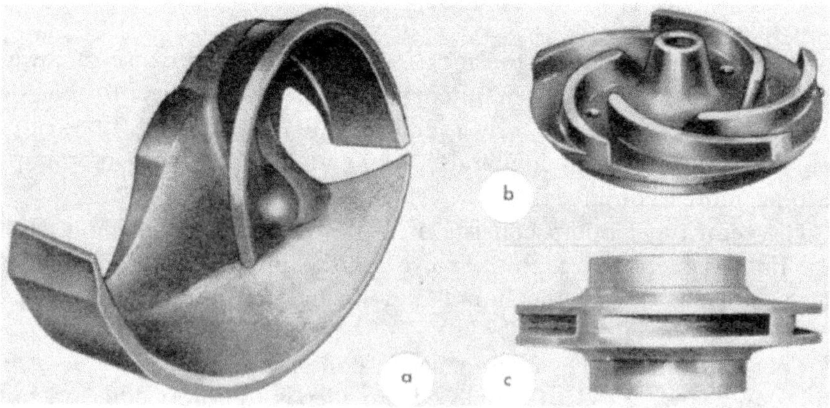

Fig. 2:4. (a) Open impeller; (b) semi-open impeller; (c) closed impeller (*Goulds Pumps*).

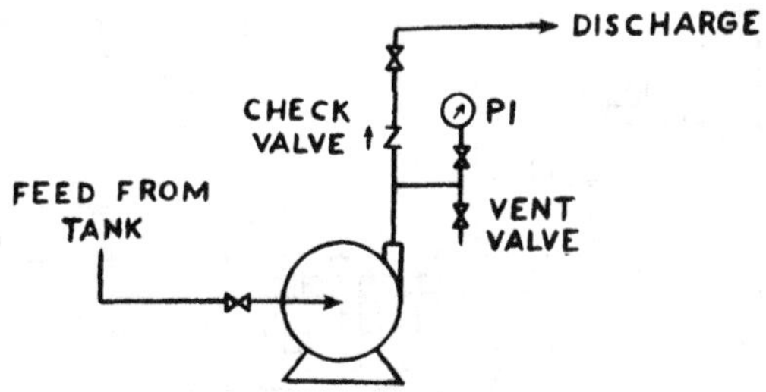

Fig. 2:5. Typical centrifugal pump piping.

shown in Figure 2 : 5.

The line from the feed tank to the pump should be as short and straight as possible and have a gate valve. If less flow is desired the globe valve on the discharge side is used to throttle the flow. The pressure gauge will indicate that the pump is putting out sufficient head. Note that the valve in front of the pressure gauge will permit removal of the gauge for testing or repair of the gauge during operat-tions; and that the vent valve (generally a needle valve) will permit samples to be taken and can also be used to vent off, or release, any gases that might be vapor-locking the pump. A check valve is sometimes placed in the discharge line to prevent liquid from backing up through the pump when the pump is turned off or accidentally stops running.

When a centrifugal pump is first installed, the flexible couplings (if used) between the shaft of the drive motor and shaft of the pump must be checked to make sure that they are in alignment. Misalignment can cause rapid wear and heating of the bearings. A screen or strainer is frequently placed in the suction line of the pump to catch any metal particles that might have been left in the tank or piping that could pass into the pump and damage it. This screen will cause a pressure drop and less efficiency, but should be used temporarily if possible. If the screen cannot be used, the tank and piping leading to the pump must be very carefully flushed.

To start a centrifugal pump, close the discharge valve, open the valves on the suction line, open the vent valve until liquid comes out, and rotate the pump by hand to see that it is free. Then start the pump and open the discharge valve to permit the desired flow. Starting the pump with the discharge valve closed protects the motor from the excessive current required for starting fully loaded. Most motors in centrifugal pumps draw more current as the rate of pumping in-

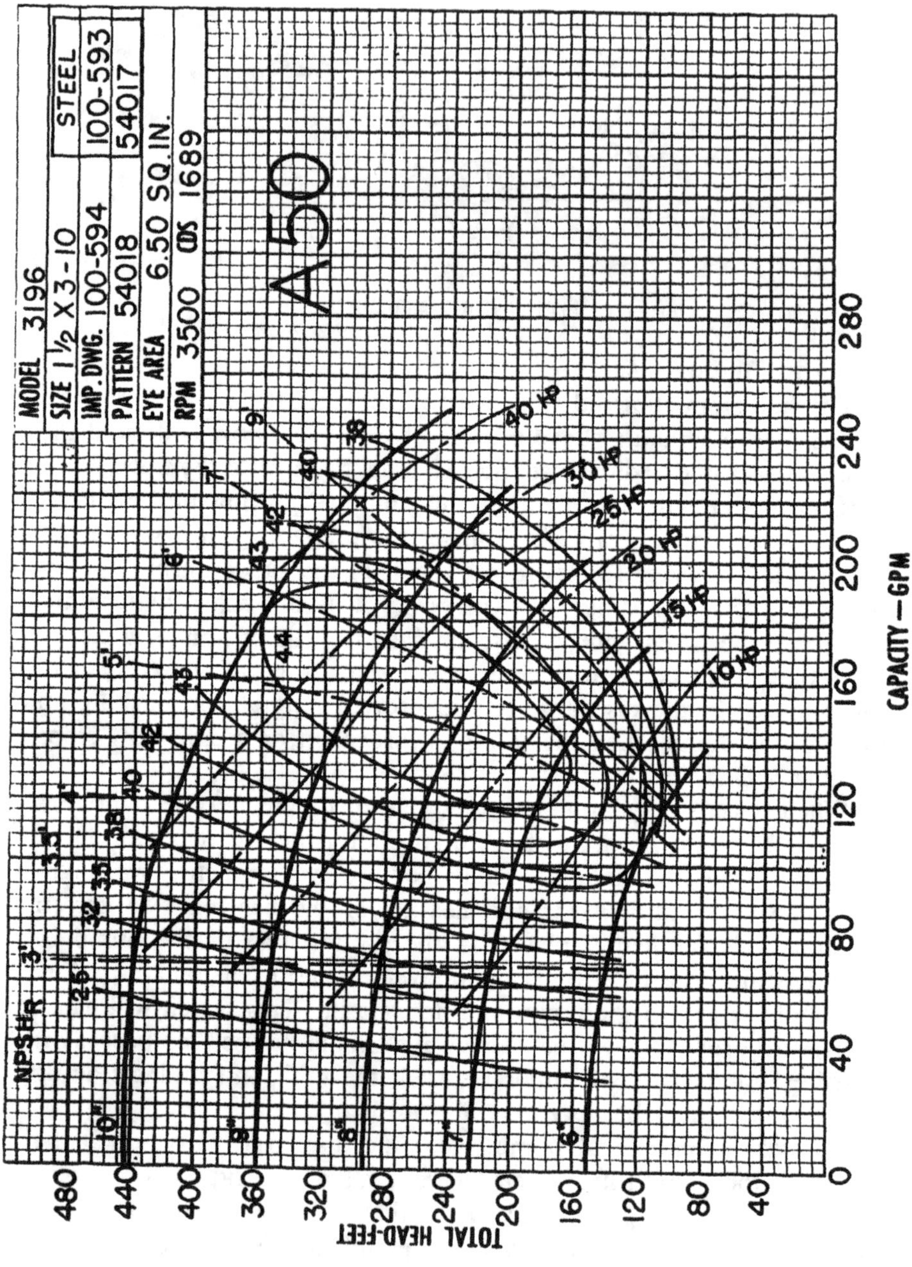

Fig. 2:6. Centrifugal pump characteristics.

creases.

If the discharge pressure is below normal for a given rate of pumping, it may be because the suction line of the pump is partially plugged or the pump is air-locked or vapor-locked.

a) If the problem is a partial plugging of the suction line, the the discharge pressure will build up when the discharge valve is closed.

b) Air-locking is more likely to occur if the feed tank has been emptied, or if the pump has been opened for repairs. The discharge valve should be closed, suction valves opened, and the air allowed to escape through the vent valve. (If a vent valve has not been provided the vent plug in the casing of the pump can be used for this purpose).

c) If the pump is vapor-locked it will generally be hot. Vapor-locking can easily happen if the pump has been running for some time with the discharge valve closed, or at a rate of flow so low that the friction of the pump has heated the material being pumped. If this is to be expected as normal operating conditions, a bypass can be provided from the discharge of the pump to the feed tank, and thus a sufficient flow can be allowed to keep the pump cool. If the pump is vapor-locked it should be turned off and allowed to cool.

Most pump manufacturers have few sizes of pump case but different sizes and types of impeller; motors of different horsepower and speed can be used to provide the necessary variations of volume or pressure. For each pump the manufacturer provides curves that will indicate the discharge pressure or head to be expected for a given rate of pumping, size of motor, and diameter of impeller. Often these curves are for pumping water and must be corrected for any other liquid to be pumped. Typical curves can be seen in Figure 2 : 6.

The curve provides all necessary data for a given pump. In this case, at a motor speed of 3,500 RPM, an eight-inch impeller with a 20-HP motor would require a NPSH of six feet and could pump up to 160 GPM of water at a discharge pressure of 230 feet of head. If the pump is required to pump over 160 GPM the motor would be overloaded because the curve of the eight-inch impeller crosses over the 20-HP curve at 160 GPH. Note that the pump size is 1½ by 3—10, which means that the pump suction connection is 1½ inches, discharge connection is 3 inches, and the maximum size of impeller is 10 inches. The curved lines numbered from 25 to 44 give the efficiency of the pump and are of interest only in the original selection of the pump.

One of the advantages of a centrifugal pump is that it has large clearances between the impellers and the case, which permits some solids to pass and permits closing the discharge valve without damaging the pump. The latter is especially important when the control valve is operated by an instrument. This same clearance, however, makes the pump less efficient than a positive displacement pump.

For a centrifugal pump with a single impeller, the pumping limit is around 300 feet of discharge head at about 500 GPM or, for heads not over 100 feet, several thousand GPM. The latter two examples would require about 100 horsepower motors. A double-impeller pump with the same motor could pump about 250 GPM at 450 feet of head, or 500 GPM at 350 feet of head. Naturally, when pumping large flows at high pressures, multiple impeller pumps are required if the horsepower of the driver is to be kept within reasonable limits.

Turbine Pumps

Turbine pumps are similar to centrifugal pumps. The power required to drive a centrifugal pump increases as the flow increases, whereas in a turbine pump the opposite is true. With a turbine pump, the discharge pressure decreases very rapidly with an increase in flow. For this reason the turbine pump is superior to the centrifugal pump where the flow rate is smaller and the discharge pressure must be higher. This type is relatively poor for liquids of high viscosity or liquids that might carry abrasives.

A drawing of the turbine is shown in Figure 2 : 7. The vanes or blades are fitted between two removable liners in the case. The outside channel is so designed that as the liquid leaves a blade it is directed

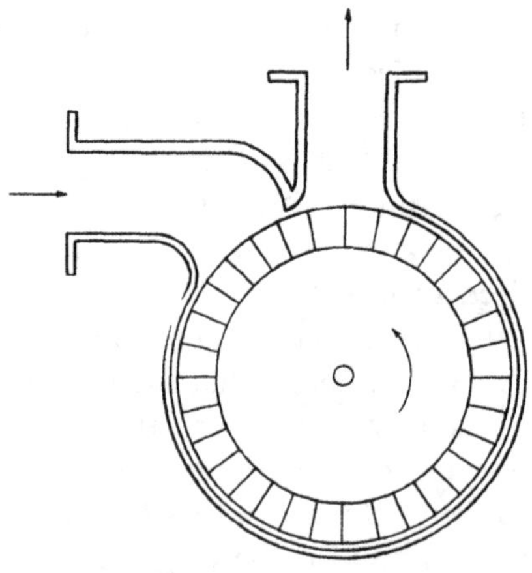

Fig. 2:7. Diagram of a turbine pump.

to the bottom of the other blades. By this method, and with the close tolerance between case and impeller, a high head or discharge pressure is developed.

Before starting a turbine pump it should be checked to make sure that all valves on the suction and discharge lines are open as shown in Figure 2 : 8. A check valve is generally placed in the discharge line to prevent gases from backing up into the pump when it is not running. If this check valve leaks, the pump must be vented. After starting the pump, the rate of flow may be throttled by using the discharge valve or by opening the bypass valve. The valves on the suction side of the pump should be gate valves; the discharge valve may be a globe valve, but the bypass valve should be a globe valve.

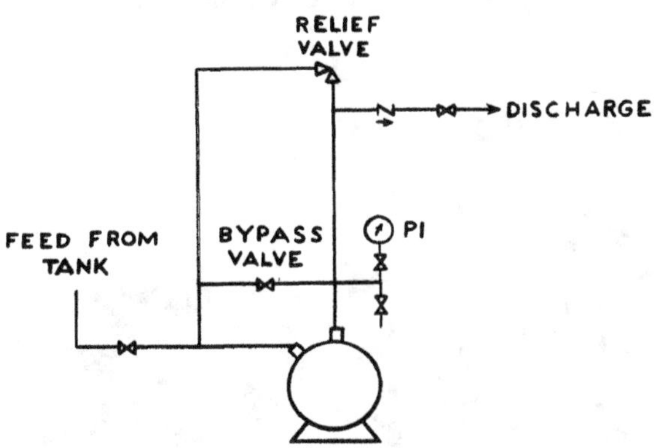

Fig. 2:8. Typical turbine pump piping.

As the pump will develop high pressures and it is difficult to leak back through a turbine pump, a safety relief valve should be provided. For the same reason, it is better to use a bypass valve to reduce output rather than throttle back on the discharge valve.

As with centrifugal pumps, more than one impeller can be used if high heads are required. The limit of a turbine pump is around 50 GPM and 350 feet of head. For example, at 50 GPM and 300 feet of head, the motor required would be about 150 horsepower; and for 7 GPM and 1,150 feet of head, it would require about 200 horsepower.

Canned Pumps

Canned pumps are so called because the rotating part of the electric motor or stator is in an enclosure similar to a can; the chemical being pumped is allowed to leak down the shaft into the electric motor. The prime feature is that this pump does not require any packing gland or mechanical seal and thus there is no leakage. This makes the canned pump excellent for use with hazardous chemicals. Actually, the pump is a centrifugal pump and it should be regarded as such for piping, starting, etc.

These pumps are not too satisfactory for use at high temperatures or with slurries. When solids may be present, the small line running from the discharge of the pump to the rear bearing will help keep the stator compartment flushed clear of solids. A line filter and cooling of the recycled material will give additional help to prevent a build-up of solids and high temperatures in the stator compartment. Figure 2 : 9 shows a cutaway of one style of canned pump.

Fig. 2:9. Cutaway of a canned pump. (*Chempump Division, Fostoria*).

Propeller Pumps

The propeller pump is similar to a centrifugal pump but it uses an *axial flow* impeller; that is, blades direct or push the flow in a direction parallel to the shaft. The blades are set at an angle to the shaft and act much like a fan (see Figure 2 : 10a). The propeller pump is good for pumping large volumes and can easily handle slurries; as in a fan, however, high pressures cannot be developed. The propeller pump is limited to about 20 feet of head. For heads from 20 to 60 feet a *mixed-flow* impeller can be used (see Figure 2 : 10b). As the type has a flared hub, the liquid receives some centrifugal force in addition to the direct axial thrust of the blades.

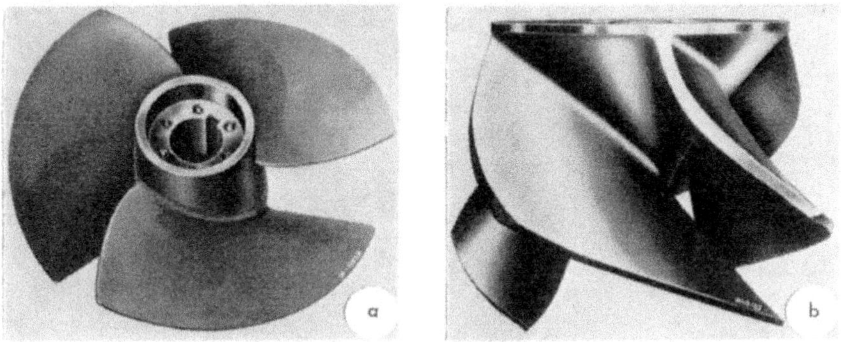

Fig. 2:10. Propeller pump.

Reciprocating Pumps

Reciprocating pumps are one type of positive displacement pump. They use a piston or plunger that moves back and forth to displace the liquid. Since they require both suction and discharge valves, this type of pump cannot be used on very viscous liquids or on liquids containing undissolved solids. The pump is excellent for developing extremely high head pressures but it is relatively expensive if large flows are desired. It is also satisfactory for delivering a fixed amount of liquid in a given time in spite of varying pressures in the tanks, temperature changes, etc.

Output of a reciprocating pump may be controlled by bypassing from the discharge side to the suction side, by varying the speed of the drive, or by adjusting the setting of the connecting rod, thus adjusting the length of stroke of the piston. Although the usual simplex (single piston) pump delivers a pulsating flow, a duplex (two-piston) or triplex (three-piston) pump can be obtained for smoother flow. In general, a surge dampener pot is also required if a relatively steady discharge pressure is required.

The piping for this pump may be similar to the arrangement previously shown for the turbine pump. Piston and plunger pumps are discussed further under Steam-Driven Pumps in this section. A cutaway of a simplex pump is shown in Figure 2 : 11.

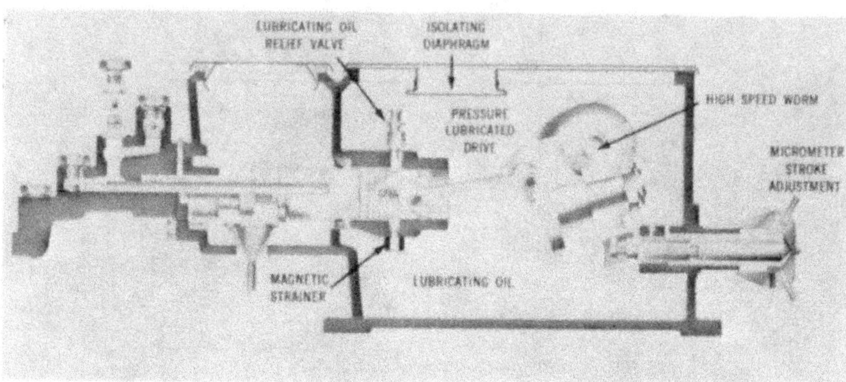

Fig. 2:11. Cutaway of a reciprocating pump (*Milton Roy*).

Diaphragm Pumps

Diaphragm pumps are almost identical to the piston pumps except that a flexible diaphragm of stainless steel, rubber, or plastic material is placed between the piston and the material being pumped, and the space between the diaphragm and piston is filled with oil or other liquid. The piston exerts pressure on the oil, which makes the diaphragm flex and puts a pressure on the chemical that is being pumped. The diaphragm seals the chemical away from the oil and piston and thus it cannot leak out. This pump is excellent for very hazardous materials where leakage of chemicals cannot be risked, but as with the piston pump it must used on clear liquids. Where breakage of the diaphragm may create a severe hazard, a second diaphragm head (called a remote head) may be used. A cutaway of a single-diaphragm pump is shown in Figure 2 : 12.

If the pump is not working, the operator should make the following checks:

1. Make certain there is no gas trapped in the chemical side of the pump by bleeding the gas out through the vent.
2. Make certain there is no air in the oil by bleeding any air out

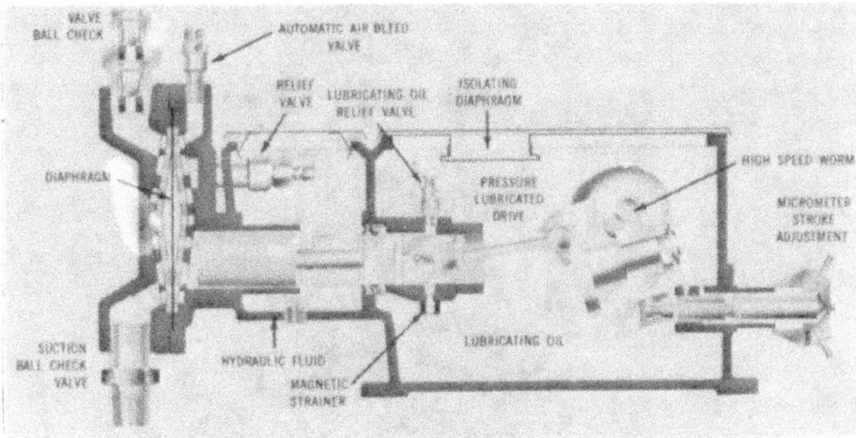

Fig. 2:12. Cutaway of a diaphragm pump (*Milton Roy*).

through the air vent in the head of the oil side of the pump.

3. If 1 and 2 do not correct the problem, reach down into the oil bath and feel the compensator valve. It should not be relieving unless the pressure rating of the pump is exceeded.
4. If 3 fails check the ball valves for dirt or solids.

The piping of a diaphragm pump should be similar to that shown for the turbine pump (see Fig. 2 : 8).

Gear Pumps

Gear pumps are another type of positive displacement pump that can deliver moderately high pressures but which must operate on liquids that are free of hard or abrasive solids. The gear pump is excellent for very viscous liquids. Generally the rate of flow is controlled by a bypass valve or by varying the speed of rotation of the gears. The piping should be similar to that for the turbine pump (Fig. 2 : 8). The style of gear pumps may vary considerably; a popular style is shown in Figure 2 : 13. As only one of the gears is driven, the other gear is free to rotate as forced by the driver.

Because of internal slippage, the gear pump is limited to about 150 *psig* on light liquids, but it may go up to 1,000 *psig* on very viscous liquids. Many gear pumps have an internal relief to prevent damage to the pump from over-pressure, but it is best to provide an external relief valve as well.

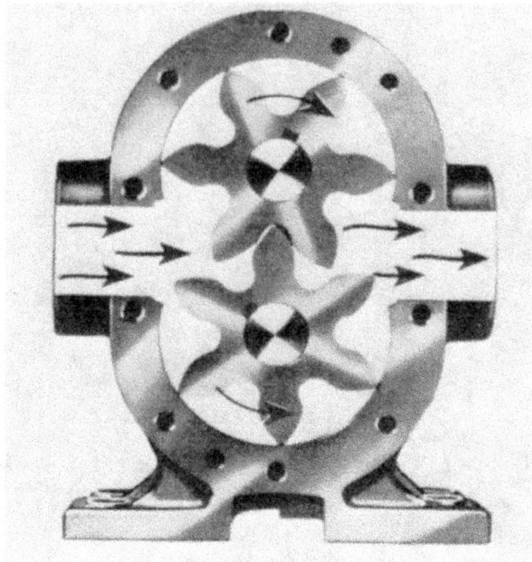

Fig. 2:13. Cutaway of a gear pump. (*Roper Hydraulics*).

Cam Pumps

The cam pump is another positive displacement type of pump. The cam is attached to the shaft and the rotor ring is free to rotate on the cam. The rotation essentially opens and closes a pocket for the liquid, thereby pumping it. This pump is best used for moderate pumping rates and discharge pressures.

Lobe Pumps

A lobe pump, shown in Figure 2 : 14, is a positive displacement type in which the two impellers are driven separately through external gears. The lobes therefore need not touch each other and a clearance of a few thousandths of an inch is allowed between lobes. (For a special use of this pump, see Section 3.)

Vane Pumps

A vane pump, another positive displacement type, consists of an elliptical casing and a cylinder with sliding vanes. The vanes are in contact with the casing and take the wear. Centrifugal force pushes the vanes outward, and as the impeller rotates, the space behind the vane first enlarges and draws in the liquid, and then decreases and forces the liquid out. The pump is used for moderate rates and pressures (see Figure 2 : 15).

Screw Pumps

Many screw pumps are essentially the same as gear pumps. There

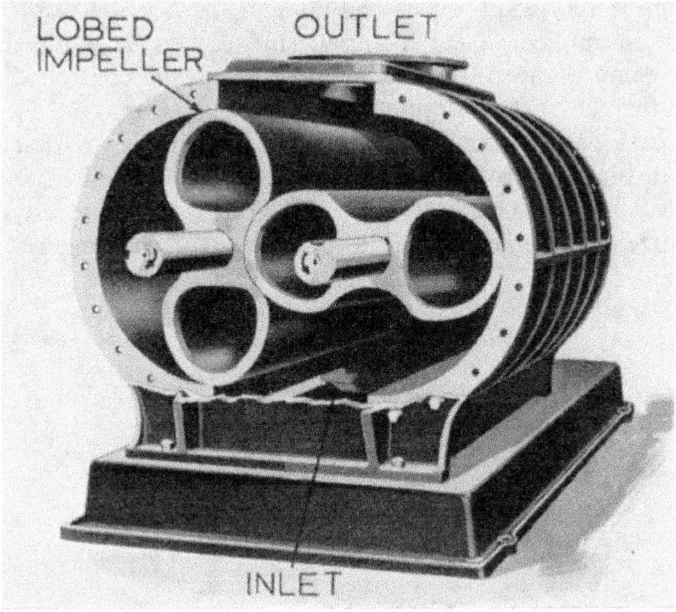

Fig. 2:14. Cutaway of lobe pump. (*Roots-Connersville Blower Division, Dresser Industries*).

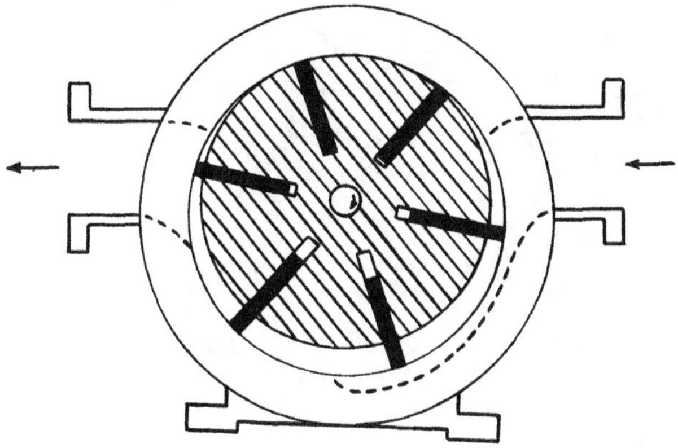

Fig. 2:15. Cutaway of a vane pump.

is one positive displacement pump, however, that consists of a helical screw that rotates inside a double helical case (Figure 2 : 16). The rotor can be obtained in rubber, stainless steel, and other materials.

This pump is very good for viscous liquids, and because of the nature of the force applied to the material, slurries can be pumped. If a slurry is being pumped, the rate of flow should be controlled by the speed of the rotor; if the liquid is free of solids, it may be controlled by a bypass valve. Generally the piping is similar to that shown for the turbine pump (Figure 2 : 8). Capacity is limited to about 250 GPM and head to 600 *psig*. The pressure developed is to a certain extent dependent upon the viscosity of the material being pumped.

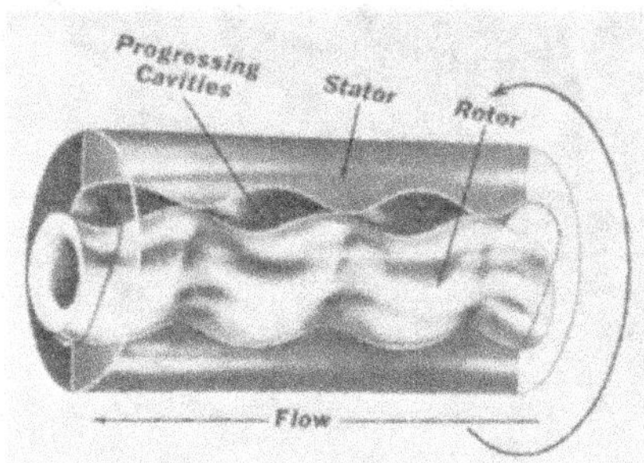

Cutaway

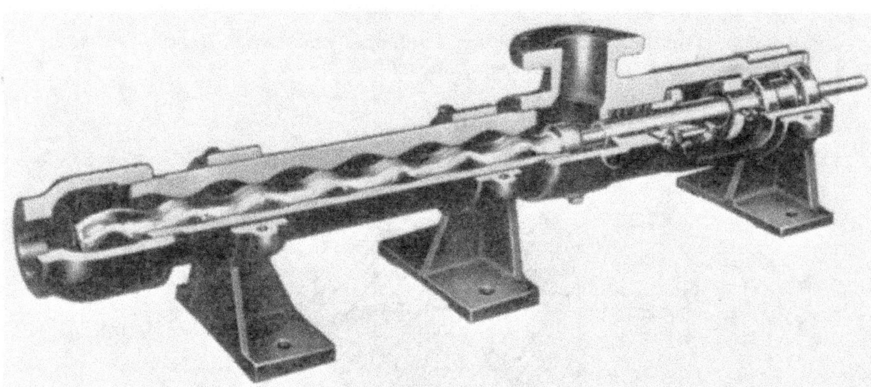

Fig. 2:16. Special form of screw pump. (*Moyno Pump Division, Robbins & Myers*).

Steam-Driven Piston Pumps

Steam-driven supply pumps for boiler water are so commonly used that a separate discussion of these pumps is justified. Details of a horizontal, steam-driven, double-acting pump are shown in Figure 2 : 17. This pump could be used for liquids other than water. The liquid end of the pump is fitted with double suction and discharge lift valves so that the pump will operate when stroking in either direction. The liquid end could be fitted with valve gears to open and close ports as shown on the steam end of the pump. The steam is used at full pressure and very little expansion of the steam occurs. The air chamber reduces the pulsations by being compressed during peak discharge and expanding slightly between peaks.

A duplex piston pump would be similar but the two pistons would give a steadier discharge pressure and steadier flow. Steam-driven pumps are generally specified in abbreviated form, for example, "$10 \times 6 \times 8$," indicating, respectively, the diameter of the steam piston, the diameter of the piston on the side of the liquid, and the length of stroke in inches.

Steam-Driven Plunger Pumps

The steam-driven plunger pump is similar to the preceding piston pump except that the packing for the piston on the liquid side is on the outside. Where the piston on the liquid is being used on high pressures this is important because external packing can be changed quickly and easily without disassembly of the pump. A cutaway of a duplex plunger pump is shown in Figure 2 : 17.

Pump Drivers

Although the eletric motor is most commonly used in the chemical

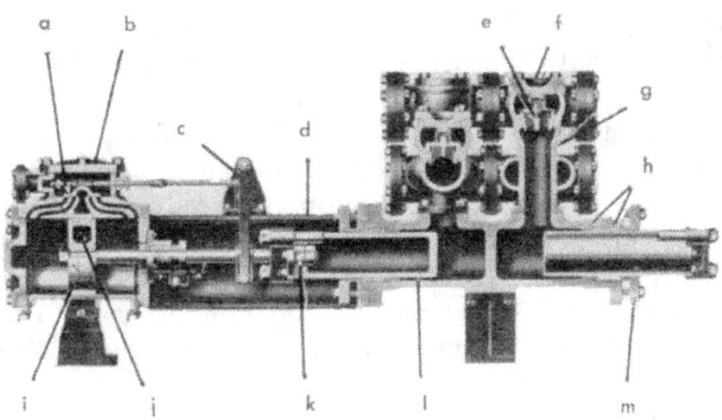

Fig. 2:17. Cutaway of a steam-driven plunger pump. (a) D-type slide valves; (b) steam chest; (c) valve motion; (d) cast iron cradle; (e) wing guided valve; (f) valve pot cover; (g) valve pot; (h) stuffing box; (i) iron piston rings; (j) box-type steam piston; (k) piston rod; (l) plunger; (m) gland.

industry, pumps driven by gasoline, natural gas, or steam engines are certainly not uncommon. When electric power fails, it is important for certain pumps to continue to function. This is why pumps that supply water to fire hydrants, for example, are frequently driven by natural gas or gasoline motors.

To keep a plant running, consideration must be given to the repair of equipment. Spares for important pumps are frequently available; that is, a second pump is standing by, completely piped and ready for operation. If trouble occurs in the pump that is being used, the operator can quickly stop it, open and close the proper valves, and start the spare pump. This keeps "down time"—non-productive time—to the minimum. It is not uncommon to find one pump driven by an electric motor and one by a steam turbine. Careful thought about what to do if power fails, or a boiler fails, is extremely important in a chemical plant.

Packing For Pump Shaft

Packing contained in a stuffing box is required on all pumps except diaphragm and canned pumps. The normal stuffing box, shown in Figure 2 : 18 (right), is designed to leak slightly and this leakage lubricates the packing. Where no leakage can be tolerated, as with hazardous chemicals or where the NPSH is so very low that air might leak into the pump, a sealing cage or lantern ring can be placed about the middle of the stuffing box to receive a sealing liquid (Figure 2 : 18, left). The sealing liquid should be a clean cool fluid supplied at a pressure slightly higher than the suction pressure of the pump. This is also a good system where the liquid being pumped contains abrasive materials or is non-lubricating. If the stuffing box of a pump is leaking excessively, the packing gland should be evenly tightened. If this does not stop the leak, the old packing must be removed and new packing inserted. Uneven tightening or excessive pressure to tighten the packing can cause excessive friction and overload the motor, shorten the life or the packing, and score the shaft. The scoring of a shaft requires complete disassembly and generally requires a new shaft.

Mechanical Seals

Mechanical seals were designed to serve the same purpose as packed stuffing boxes. When they are installed by a properly trained mechanic they will cause less frequent "down time" than packed stuffing boxes. A single seal is shown in Figure 2 : 19, but double mechanical seals are also available in which a clean cool liquid is

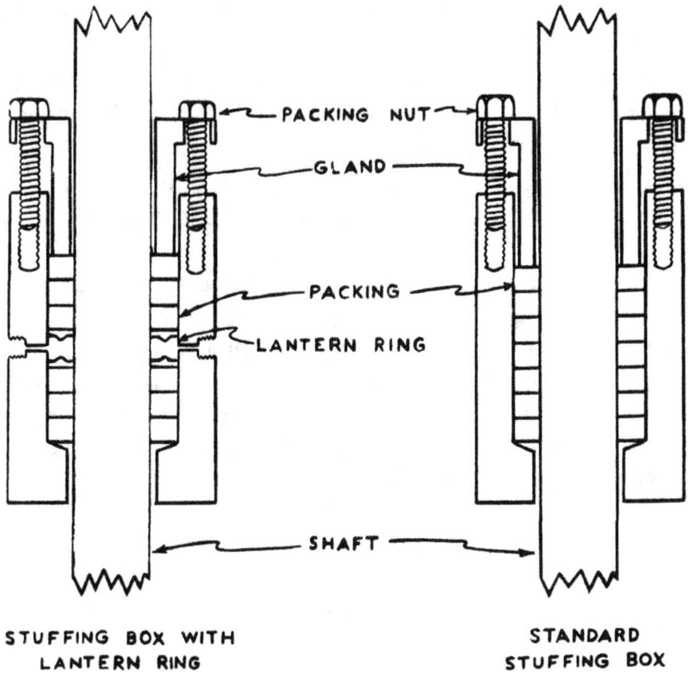

Fig. 2:18. Stuffing box.

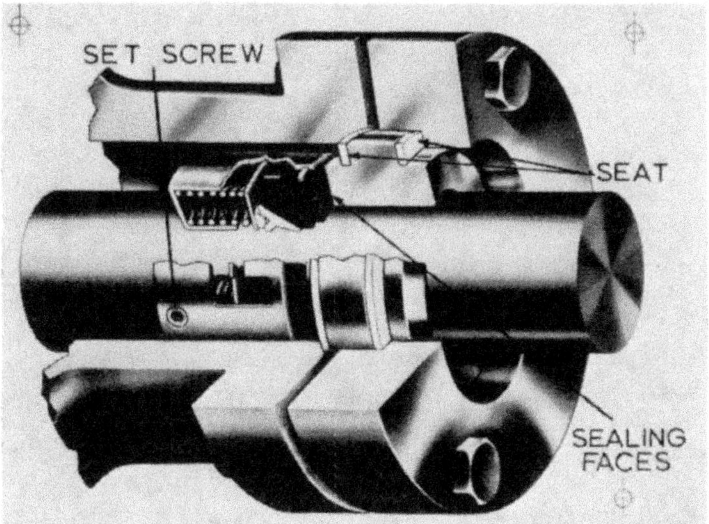

Fig. 2:19. Cutaway of a single mechanical seal. (*Crane Packing*).

supplied between the two units at a pressure slightly greater than the suction pressure of the pump. Double seals should be used where no leakage can be tolerated, or for slurries. The seals depend upon a sliding contact and the sliding faces are held together by springs. The other two effective surfaces are the Teflon wedge, which holds the rotating part of the seal against the shaft, and the two Teflon seat gaskets, which hold the stationary part of the seal to the stuffing box.

If the seal has not been operated for some time and there is a chance that the sliding faces might be stuck together, the shaft should be rotated by hand to loosen the faces and thus safeguard them from damage due to the sudden powerful jerk of the motor. A single seal depends upon some leakage to lubricate the faces and help the sealing. Where double seals are used, the supply of sealing liquid should be operating before the pump is started.

3. GASES—STORAGE AND HANDLING

GENERAL

Small quantities of gases are generally stored in high-pressure standard cylinders. One of the smaller sizes has a nominal capacity of 80 standard cubic feet of gas and measures 6-5/8 inches inner diameter by 38 inches in length. Two other common sizes are 110 cubic feet and 220 cubic feet. The largest commonly used cylinder is 300 standard cubic feet of nominal gas capacity and measures 8-3/4 inches inner diameter by 61 inches in length. (Length does not include valve and cap).

As the cylinders may be used for various gases, the valve outlet connection is so constructed as to permit only the proper reducing regulator to be attached.

Although a number of these cylinders can be *manifolded*—that is, joined together by piping for larger quantities of gas—there is a practical limit. Where quantities of about 500,000 standard cubic feet must be stored, a sphere about 50 feet in diameter, designed for gas pressure of about 50 *psig*, would be a practical answer.

For very large quantities of gas, the telescoping gas holder is frequently used. This consists of a bottom container with an open top into which is fitted another container with an open bottom. A seal, generally liquid, is maintained between the inner and outer containers. The top container rides up and down in guides and floats on the gas pressure; thus the whole container increases or decreases in volume as needed. This telescoping gas holder is generally used at an operating pressure of 2 or 3 *psig*, and is commonly used as a natural gas surge tank.

HANDLING AIR

For moving air, the common fan is most economical; however,

the discharge pressure is generally limited to about one or two *psig*. A single-stage rotary compressor will move up to 3,000 standard cubic feet a minute (SCFM) at discharge pressures to 50 *psig*; and the two-stage rotary compressor will move about 2,000 SCFM at discharge pressures to 250 *psig*. The single-stage centrifugal compressors will move about 60,000 SCFM with discharge pressures to 8 *psig*; and multistage units can go to 200,000 SCFM at discharge pressures to 5,000 *psig*. For large volumes of air the axial compressor can be used to move up to 1 million SCFM at discharge pressures to 100 *psig*. For moderate volumes of air at pressures over 50 *psig*, the piston type compressor is generally used.

To remove air in order to create a vacuum, the rotary vacuum pump or steam jet is generally used.

TYPES OF APPARATUS

Axial Compressors

The axial compressor can be considered as a series of fans. Numerous blades are set on a horizontal shaft that, turning at high speed, compresses and moves the air. It is an excellent compressor for large volumes of air where high pressure is not required. Generally, it is limited to about 200,000 SCFM and 50 *psig*, although about 125 *psig* is possible.

Fans

For moving large volumes of air at pressures below one *psig*, the fan is very economical. Horsepower requirements and noise level increase very rapidly with increase in pressure requirements.

Reciprocal Compressors

For smaller volumes of air at higher pressures, the reciprocal (or piston) compressor is excellent, and is probably the most commonly used of all the gas compressors. A typical installation of a reciprocal compressor for supplying clean dry air to instruments is shown in Figure 3 : 1.

If the air is compressed to 50 *psig*, the temperature of the air will be about 300°F. An *after-cooler* is provided to cool the air and thus remove most of the moisture. A surge tank is provided for air storage. The pressure in this tank is detected by a pressure-controller (not shown on the drawing), regulating the operation of the compressor.

For use in instruments, the air is generally run through dryers, e.g., silica gel units, so that moisture will not collect in the instruments. In order to protect the compressor from serious damage,

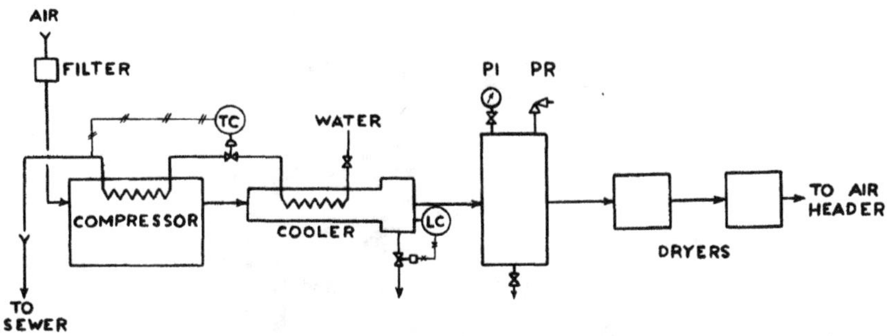

Fig. 3:1. Typical reciprocal compressor system.

switches (not shown in the drawing) are provided to stop the compressor if the pressure of lubricating oil is too low or if the temperature of the compressor itself goes too high.

Rotary Blowers and Compressors

The cam, lobe, and vane pumps discussed in Section 2 on handling liquids can also be used to move gases. The lobe pump is frequently used to create the vacuum on rotary vacuum filters or distillation columns but a water seal is introduced. Most of the rotary units are good for moderate volumes of a few thousand SCFM at maximum discharge pressures of 25 to 50 *psig*, and for service with vacuum. Special units are available that utilize more than one stage of compression for higher discharge pressures to about 250 *psig*.

Vacuum Pumps

The rotary pumps referred to above can be used for vacuum service where a moderate vacuum at moderate air rates is required. The maximum practical vacuum is in the range of 10 to 40 mm. Where larger volumes of air are handled at moderate-to-high vacuum, steam jets (discussed later) are very good. A three-stage steam jet can be used to about 2 mm, and a four-stage, to about 0.5 mm. For higher vacuum—to 0.00000001 mm—a rotary pump assisted by an oil-sealed pump is generally used. The oil-sealed pump pulls the vapor and compresses it to 0.05—0.2 mm, and the rotary pump compresses this to atmospheric pressure to discharge the vapors.

One of the high-vacuum types of pump is a rotary oil-sealed pump. Referring to Figure 3 : 2, the air is drawn in and compressed as in a cam pump. The oil acts as a seal to enable the pump to operate at the high vacuum. Air and entrained oil leave the compression

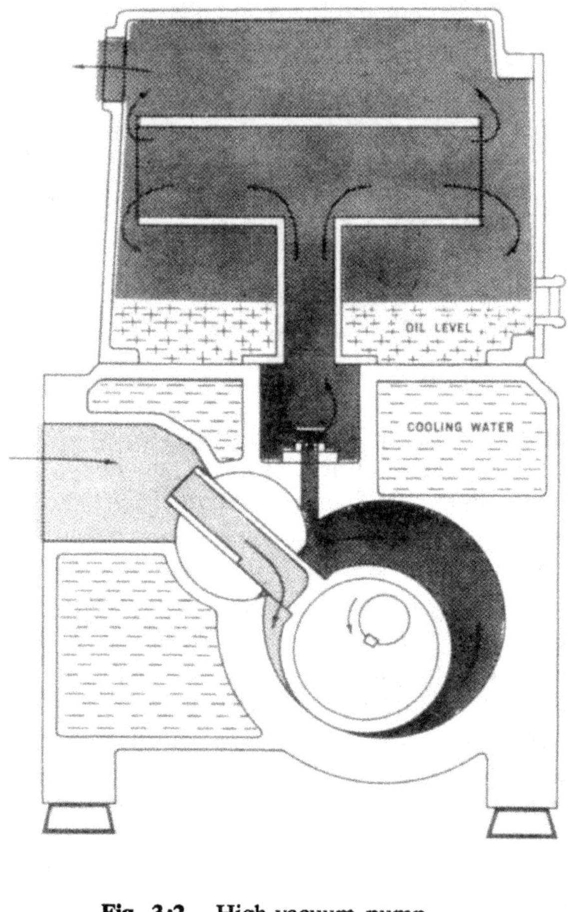

Fig. 3:2. High-vacuum pump.

chamber through a so called *poppet valve* (see Section 4) and enter a separator.

The heat of compression is carried away by a cooling jacket. As it is necessary to prevent solvents and water from collecting in the oil, a cold trap must be placed between the vessel at low vacuum and the vacuum pump.

The high-vacuum pump can handle vacua to about 10 microns, or 0.01 mm of mercury, but above this it should be used in conjunction with a rotary pump as previously mentioned.

Steam Jets

As discussed under Vacuum Pumps in Section 2, the steam jets are commonly used to move large quantities of air to provide vacuum to 0.5 mm of mercury. The steam jets are frequently referred to as *ejectors*. A cutaway of a steam jet is shown in Figure 3 : 3. The steam goes through a nozzle that reduces the steam pressure and uses this power to create a high velocity. The high-velocity stream pulls the gas or fluid from the opening, which is at a right angle to the high-velocity flow of steam. The mixture then goes to the diffuser where the residual velocity is used again to increase the pressure for discharge.

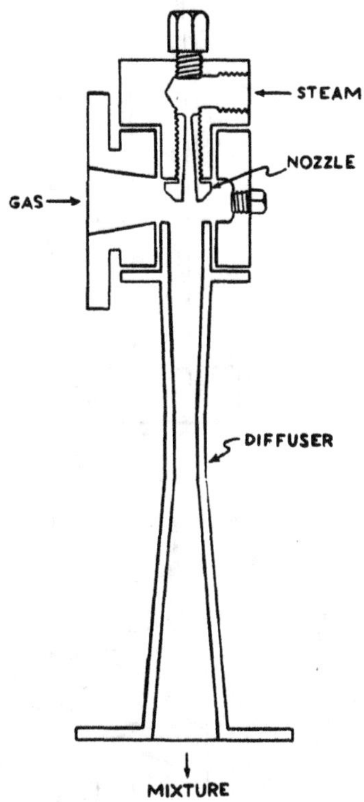

Fig. 3:3. Cutaway of steam jet.

A single steam jet can create a vacuum to 3—4 inches of mercury (about 75 to 100 mm of mercury) using 100 *psig* of steam. If steam of sufficient pressure is not available, two stages are used to achieve the same result. Two stages with a condenser can be used for vacau to 25 mm. A typical installation is shown in Figure 3 : 4.

The inter-condenser condenses the steam of the high-vacuum

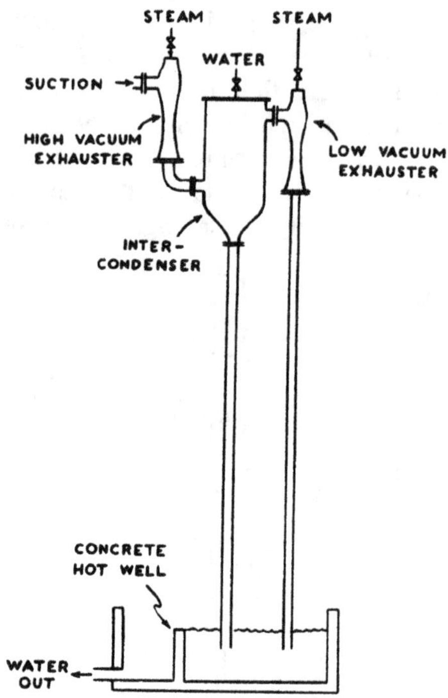

Fig. 3:4. Two-stage steam jet.

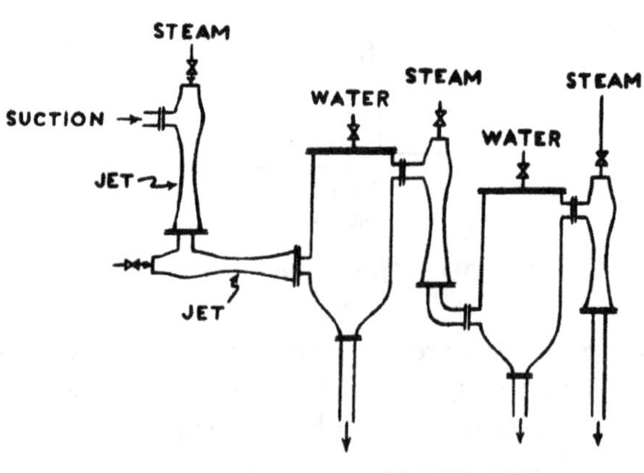

Fig. 3:5. Typical four-jet system.

exhauster or jet so that only the non-condensable gases, which have been cooled to a few degrees above the water temperature, reach the low-vacuum exhauster. The tail pipe, or barometric leg, running to the hot well must be 34 feet long because the low-vacuum exhauster pulls a vacuum on the inter-condenser. If a barometric leg or tail pipe is impractical because of the height of the building, a centrifugal pump can be used instead of the leg and hot well; but generally a barometric leg is preferred.

A three-stage system will provide a vacuum to 0.5 mm of mercury, and a four-stage unit could be used primarily for greater steam efficiency. A typical four-stage system is shown in Figure 3 : 5.

A steam jet is designed to operate with steam supply at a certain pressure. If the pressure of the steam supply drops even a little, the vacuum and the performance of the jet will be reduced. It is also important that sufficient water be supplied to the inter-condenser in order to condense the steam properly.

4. PIPE, FITTINGS, VALVES

PIPE

Pipe used in the chemical industry today can be made of steel, steel alloys, cast iron, vitrified tile, lead, glass, glass- or plastic-lined steel, plastic, aluminum, copper, or wood. It is not uncommon to find almost all of these materials used in the same plant. The pipe chosen is generally the most economical of those materials that will be resistant to the corrosive effect of the chemicals being handled. Pipe of the proper size and wall thickness is then selected on the basis of maximum pressure drop allowable, temperature, and pressure.

Steel

The most commonly used pipe is of steel, commonly referred to as *carbon steel* or *black iron*. Up to 12 inches size of pipe means the nominal inside diameter of the pipe; over 12 inches, the size is the nominal outside diameter.

Wall thickness at one time was designated as *standard, extra strong,* and *double extra strong*, but today wall thickness is identified by *schedule number*. Eleven of these schedule numbers have been adopted: 5, 10, 20, 30, 40, 60, 80, 100, 120, 140, and 160.

In pipe up to and including 12 inches in diameter, the standard pipe that is most often used is schedule 40; schedule 80 is probably the next most popular. In one-inch pipe of low-carbon steel for example, of which the outside diameter is 1.315 inches, in schedule 40 the wall is 0.133 inch thick, whereas in schedule 80 the wall is 0.179 inch thick. Schedule 160 is seen in some high-pressure plants but the remaining schedule numbers are not too popular. Most plants try to limit the number of schedules used in order to avoid confusion and excessive inventories. In small sizes, one inch and below, most plants do not use any wall thinner than schedule 40.

Stainless Steel

Stainless steel pipe is popular in the chemical industry because of its general resistance to corrosion. Stainless steel is classified by type number. Of the 300 series, which shows good general resistance to corrosion, the most popular is type 316 with type 304 being next in order. Type 304 is more economical than type 316 but the general resistance to corrosion is less. The 300 series has high chromium and nickel content. The major difference between the two is that type 316 has 16 to 18 percent chromium, 10 to 14 percent nickel and 2 to 3 percent molybdenum, whereas type 304 has 18 to 20 percent chromium, 8 to 10 percent nickel, and 2 percent manganese.

The 400 series of stainless steel is more economical than the 300 series, but the general resistance to corrosion is less. The 400 series is not too commonly found; type 410 is probably the most popular of this series. Type 410 has 11.5 to 13.5 percent chromium, and the higher the number in the series the higher the chromium content. A good way to tell the 400 series from the 300 series is that the 300 series does not have magnetic properties.

Most plants limit the types of stainless steels used in order to avoid confusion and excessive inventories. The pipe most commonly used is schedule 40, with schedules 5 and 10 next in order. Schedules 5 and 10 are often used for low-pressure ducts or where corrosion is is not expected. As it is rather difficult to cut threads on stainless steel and ensure tight connections, stainless pipe is most often welded, or screwed and back-welded. When back-welding is planned. plastic sealing compounds, generally referred to as "pipe dope" should not be used. Schedules 5 and 10 have such thin walls that they must be welded.

Copper

Copper tubing is extensively used in most plants to connect instruments to the air supply and to carry air signals from one instrument to another. Generally this is soft copper, purchased in 50 foot rolls.

Glass And Lined Steel

Glass pipe is resistant to the corrosive effects of almost all chemicals and has the advantage of permitting one to see what is going on in the pipe. The present glass pipe can withstand considerable pressure in the smaller sizes of pipe. Glass pipe is relatively expensive and must always be securely supported to prevent vibrations that would cause leakage of the flanged joints.

Steel pipe with a lining of glass, Saran, Kel-F, or Teflon, has all the advantages of glass pipe except transparency. The lined steel pipe is much stronger than the glass pipe and in pipe of larger diameter this becomes very significant. This type of pipe is very expensive when compared with others but it is extremely good in every way. Valves and fittings similarly lined are available.

Plastic

Plastic pipe is available and some types are very economical. The low-priced plastic pipe is quite resistant to most acids but is softened and ruined by many solvents. The temperature is generally limited to about 150°F for the lower-priced pipe and the piping must be supported throughout the entire length by, for example, steel channel.

Plastic tubing is being used more and more as a substitute for copper tubing, for the purposes mentioned for the latter material.

FITTINGS

Once the pipe has been selected, the choice of fittings is next in importance.

Iron

Screwed fittings for steel pipe are available in cast iron; 25 lb (light), 125 lb (standard), 250 lb (heavy). They are also available in malleable iron, 150 lb. Fittings for welding are available in types for both butt welding and socket welding, and come in *light*, *standrard*, and *double extra strong* weights.

These *standard* fittings are commonly used with schedule 40 pipe; the *extra heavy* correspond to schedule 80 pipe. Black iron pipe up to two inches in diameter is generally screwed. From two inches to about twelve inches in diameter it may be screwed or welded, but over twelve inches it is generally welded.

Stainless Steel

Stainless steel fittings similar to the cast iron fittings mentioned are available and are generally purchased in the same type of stainless steel as the pipe on which they are to be used. As it is difficult to thread stainless steel pipe and obtain a tight connection, it is generally welded, or screwed and back-welded. The use of Teflon tape over the threads has helped greatly in securing tight screwed joints.

As it is difficult to thread both ends of short lengths of pipe, these short lengths called *nipples* are generally factory-made and purchased in lengths to about six inches. There are many special fittings but the common ones are shown in Figures 4 : 1 and 4 : 2.

Welded pipe is generally joined to valves, tanks, pumps, etc., by

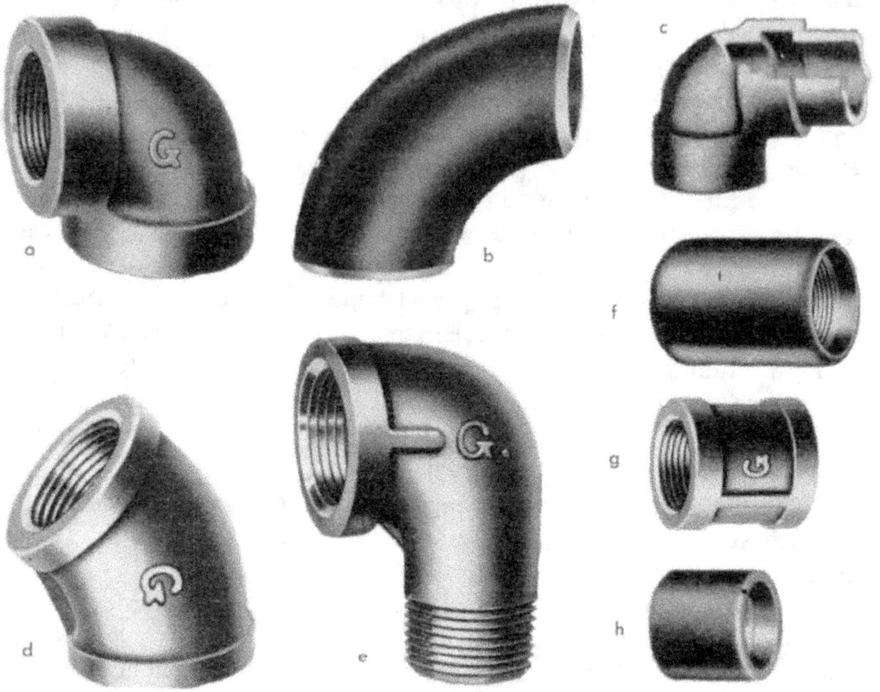

Fig. 4:1. Various ells and couplings. (a) 90° Screw elbow (*Grinnell*); (b) 90° butt weld elbow (*Grinnell*); (c) 90° socket weld elbow (*Vogt*); (d) 45° screw elbow (*Grinnell*); (e) 90° street elbow (*Grinnell*); (f) wrought coupling (*Grinnell*); (g) cast coupling (*Grinnell*); (h) socket weld coupling. (*Vogt*).

flanges. Flanges are specified by the size of pipe on which they are to be used, by the pressure expected in the pipe, by materials of construction, and by type of face. For example, "a 150-lb, type 316 stainless steel, slip on, raised face, one-inch flange" would be an adequate description. The specification of 150-lb determines, among other things, the location of the bolt holes, which is important because a 150-lb flange cannot be bolted to a 300-1b flange. Generally the faces must also match. The common methods of attachment to the pipe and common faces are shown in Figure 4 : 3. Any combination may be obtained, i.e., a flat face flange may be *slip on*, or *butt weld*, and so on.

A number of other faces available are similar to the ring joint type

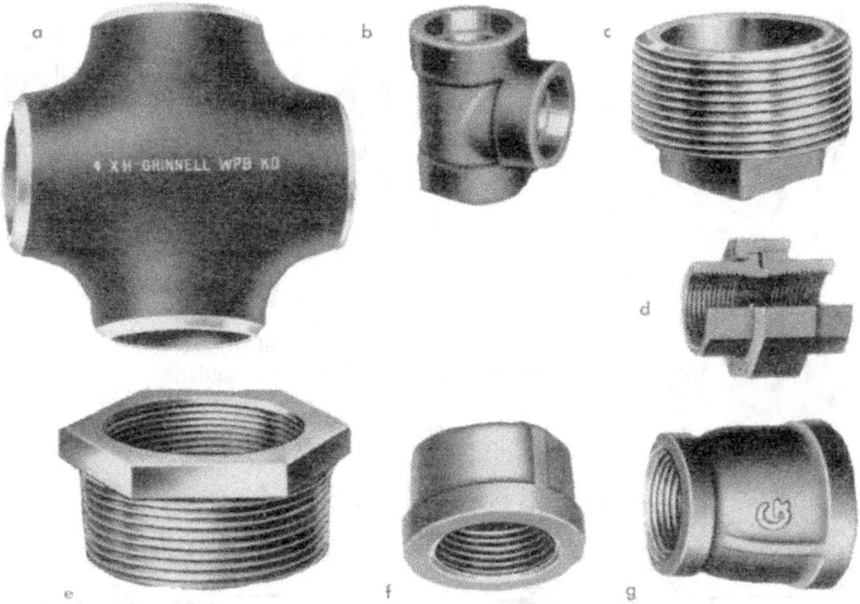

Fig. 4:2. Miscellaneous Fittings. (a) Butt weld cross (*Grinnell*); (b) socket weld tee (*Vogt*); (c) reducing bushing (*Grinnell*); (d) female screw union (*Grinnell*); (e) square head plug (*Grinnell*); (f) standard cap (*Grinnell*); (g) standard concentric reducer (*Grinnell*).

(Figure 4 : 3c). The most popular are *tongue and groove* and *knife edge*.

Copper

Copper tubing used in chemical plants to connect instruments is generally joined by *compression* or *flared* fittings. Details of these two types are shown in the cutaways in Figure 4 : 4. A flaring tool must be used to shape the tubing in order to use the flared fittings.

These fittings are available in 1/4-, 3/8-, 1/2-, and 1-inch sizes: larger sizes are occasionally seen but 1/4-inch is sufficient for use with most instruments. The fittings are available in tubing-to-tubing *unions*, *ells*, *tees*, etc., and are also available in pipe-to-tubing ells and connectors with both male and female pipe ends.

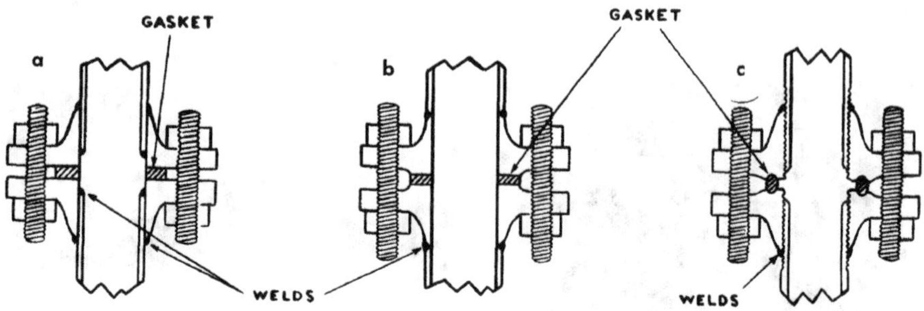

Fig. 4:3. Cutaways of various flanged joints. (a) Flat or full face—slip-on flange welded front and back; (b) raised face—welded-neck flange butt welded; (c) ring joint—screwed flange back welded.

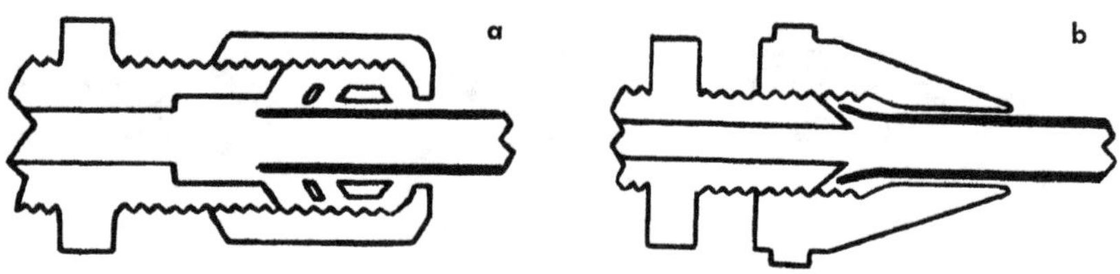

Fig. 4:4. (a) Cutaway of a flared fitting; (b) cutaway of a compression fitting.

Glass

Glass pipe fittings are purchased in the shapes desired, such as *tees*, *reducers*, *ells*, etc., and are fastened using an envelope-style gasket of Teflon and metal flanges. Springs placed under the bolts of the flange help to prevent uneven tightening with resultant extra stresses (Figure 4 : 5).

Plastic

A full line of plastic fittings is available to match most piping. Both screwed and socket-type fittings may be used with many plastic pipes. For some plastics a quick-setting adhesive can be obtained to cement the fittings to the pipe. Plastic tubing for air lines in instruments can use compression fittings similar to those described under *Copper*.

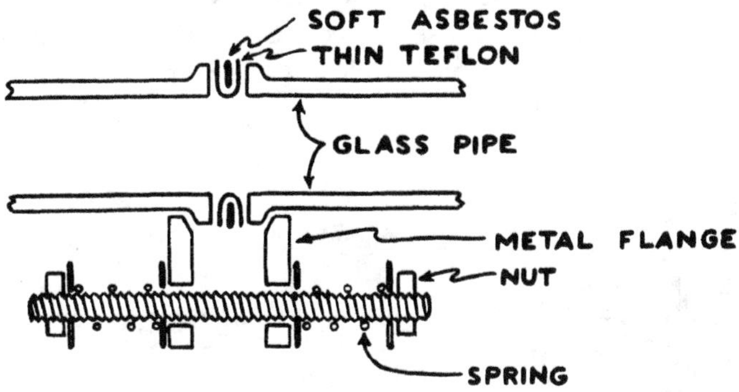

Fig. 4:5. Cutaway of a glass pipe joint.

Lined Pipe Fittings

Plastic- or glass-lined pipe fittings are available to match the pipe. Generally these are joined by flanges where the face of the flange is covered with the liner and a gasket of Teflon or Kel-F is used. Manufacturers' catalogues show the pipe, fittings, and valves that are available. Diaphragm-type valves are generally used with lined pipe.

POINTERS ON PIPING

Screwed Pipe

In cutting threads, the die must be sharp and properly set or the joints, will not be tight. For 1/2-inch pipe the threads should be cut for a distance of one-half inch, a few more threads for 3/4-inch pipe, and 3/4-inch of threads for one-inch pipe. Three-inch pipe requires about one inch of threads, and six-inch pipe about 1-1/4 inches of thread. Either too little or too much will give a joint that might very well leak.

Make sure the threads are clean, using a wire brush, and apply the proper pipe dope or tape to the male threads only, i.e., on the piece to be inserted. Metal chips and pipe dope inside the pipe can cause a serious amount of damage to the valves.

Be very careful of the threads so as not to damage them. If pipe is not to be used immediately, use plastic or metal snap-on pipe-caps or plugs, if they are available, to protect the threads and keep the pipe clean.

Wrenches

The proper size and type of pipe wrench is very important, and reinforcing the wrench with a vise or second wrench generally referred to as a "back up" wrench, placed close to the first one, is also very important (Figure 4 : 6).

Fig. 4:6. Showing proper wrench use (a); dangerously large wrench being used (b).

On a hexagonal head use a "hex" (smooth jaw) or monkey wrench (Figure 4 : 7a). If one is not available the equipment can be protected by softening the jaws on a pipe wrench. The jaws may be made

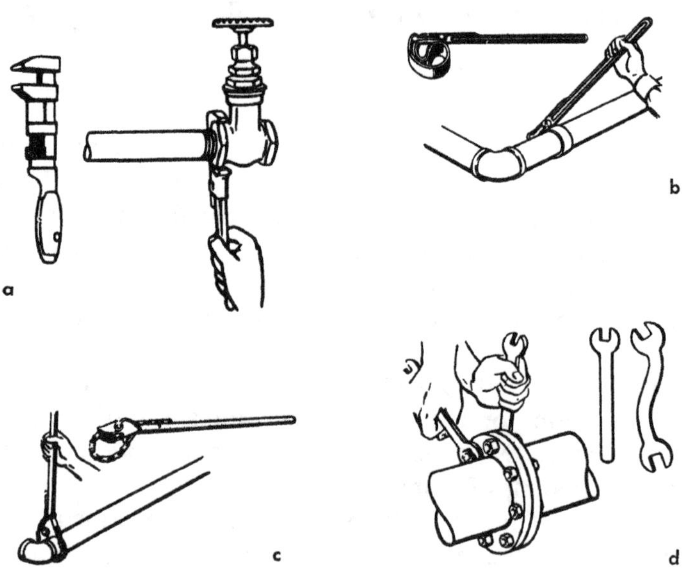

Fig. 4:7. Showing use of proper wrenches.

"soft" by using pieces of sheet lead to cover the teeth of the wrench.

The *strap wrench* (Figure 4 : 7b) is generally used on shafts and pipe that must be protected from scratches or nicks. The *chain wrench* (Figure 4 : 7c) is usually reserved for pipe sizes three inches and over.

Flanged Pipe

Flanged pipe should be aligned and properly supported before drawing up on the flanges. Bolts should be lubricated and nuts run in by hand. The gasket should always be properly centered. Open-end or socket wrenches (Figure 4 : 7d) should be used to tighten the bolts successively in the order shown in Figure 4 : 8, to insure proper alignment and a good tight flange.

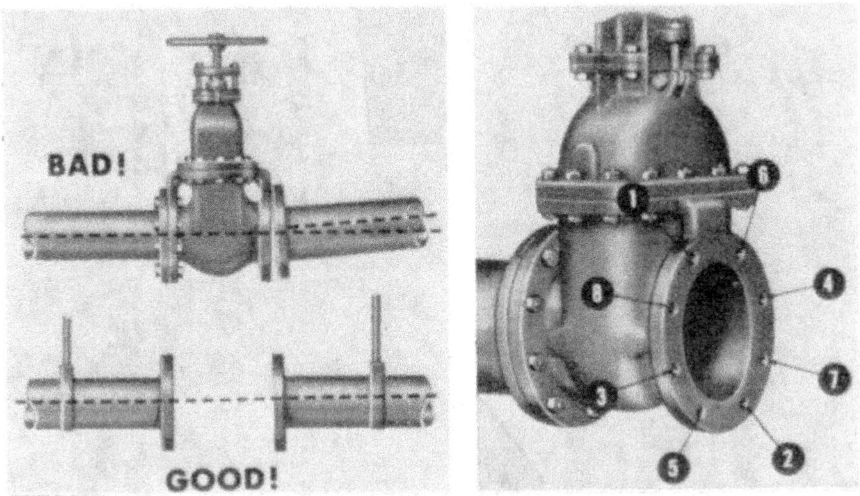

Fig. 4:8. Showing pipe aligning and bolt tightening sequence.

VALVES

Gate Valves

Gate valves have a gate, disc, or wedge that can be raised or lowered. As the drop in pressure across a fully open gate valve is very low, this is an excellent valve for stopping flow but not for throttling it. A partially open gate would have high velocity of flow at right angles to the gate. This would cause vibration and erosion of the machined faces of the disc and prevent the disc from proper seating to stop the flow completely. In addition, the shape of the opening as the valve is opened is not suitable for controlling flow. The first few turns give very little opening, but when the valve is half-open a few turns change the size of the opening considerably. The seriousness of the damage to a gate valve resulting from using it in a partially open position cannot be over-emphasized. Most gate valves have a wedge-shaped disc with matching seats and the refacing or repairing of these seating surfaces is not a simple operation.

Gate valves may be obtained for flanged, screwed, or welded connections. The bonnet may be screwed or bolted and the size, pressure

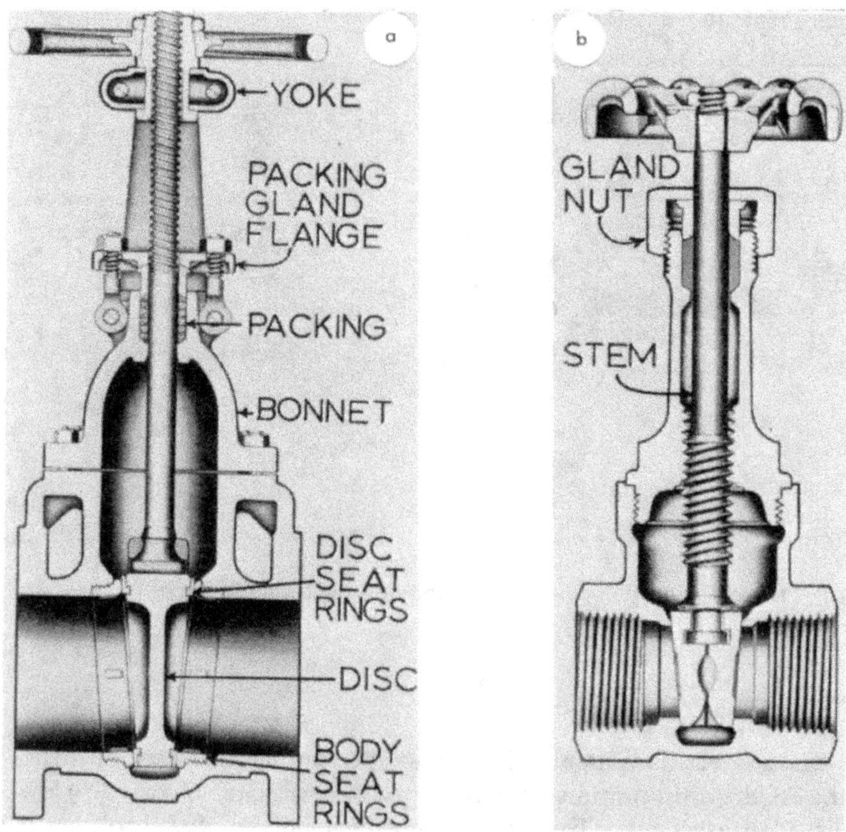

Fig. 4:9. Cutaway of two styles of gate valve. (a) flanged valve with solid wedge disc, rising outside screw stem and bolted bonnet; (b) screwed valve with split-wedge disc, rising inside screw stem and screwed bonnet (*Crane*).

ratings, and materials of construction may vary. The valve generally has a rating, such as 125, 150, 300, etc., which refers to the steam pressure for which it is rated. If the valve is to be used on low-temperature chemicals it will be suitable for higher pressures than the steam rating. Sometimes the valve will have two ratings on the body, such as 125 *S*-200 *WOG*, which means that it is designed for 125-lb steam service or 200-lb service of water, oil, or gas. The manufacturers' table on service recommendations should be consulted before using a valve because both the temperature and the pressure at which the valve is to be used are important. Figure 4 : 9 shows two of the many styles of gate valves.

Globe Valves

In the globe valve, the direction of flow is changed so that it runs essentially parallel to the seating surfaces. This reduces erosion of the seating surfaces, but by changing direction of the flow several times the pressure drop through the valve is considerable. The globe valve is designed to throttle or control the flow. The opening is so designed that if four turns of the stem are required for full opening or closing of the valve it will permit about one-quarter of the flow to pass for each turn of the stem.

Both the disc and seat of the globe valve are designed for relatively easy renewal. The disc can be a plug of metal making contact with a metal seat or it can be made of a composition. Like the gate valve, a globe valve may be obtained in various styles, sizes, and pressure ratings. Two of the many styles available are shown in Figure 4 : 10.

Needle Valves

The needle valve is similar to the globe valve except that the end of the stem is tapered to a point instead of ending in a disc. This furnishes precise control and is excellent where a small stream is desired and a large drop in pressure is available. One common application of this valve is in sampling, especially where the sample is being taken from a line under high pressure. One style of needle valve is shown in Figure 4 : 11.

Plugcock or Ball Valves

The plugcock or ball valve is similar to the gate valve except that the opening is a hole in a ball- or cone-shaped plug that rotates to stop the flow or let the flow pass through the hole. As with the gate valve, there is very little drop in pressure through these valves when they are fully open. Whereas solid material might gather and hinder the movement of the opening in a gate valve, there is no place

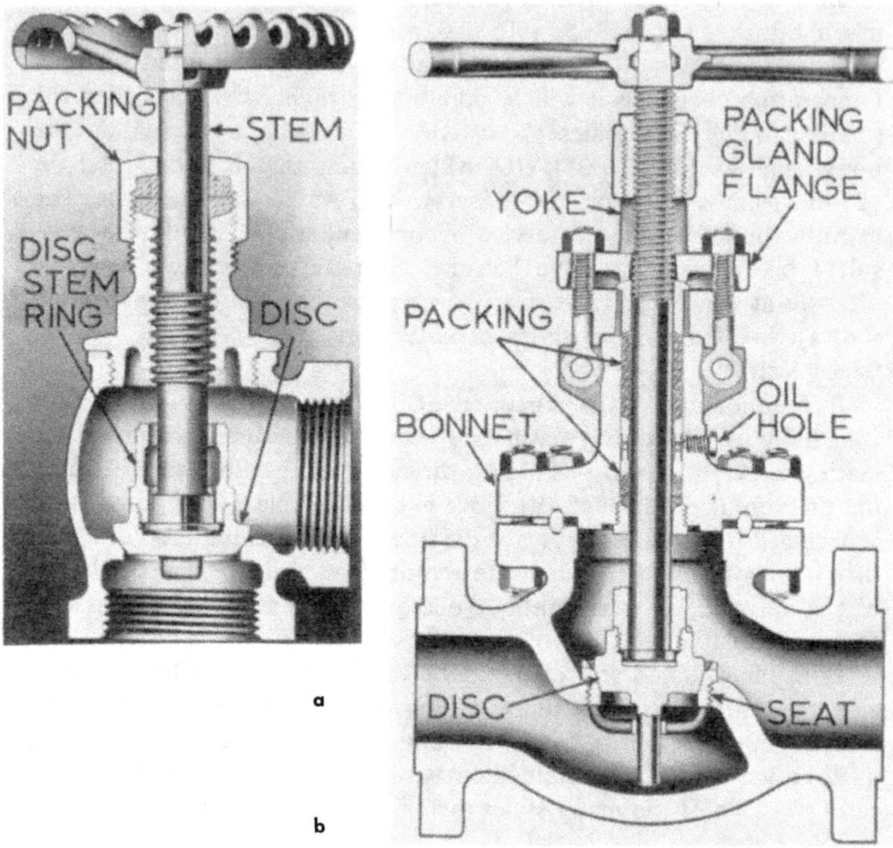

Fig. 4:10. Cutaways of two styles of globe valve. (a) screwed angle valve with screwed bonnet and packing gland. (b) flanged valve with outside screw and yoke (*Crane*).

where it could gather to hinder movement of a rotating ball or plug. These valves are therefore good for service with a slurry. Because of the large area of sliding surface, these valves have a tendency to stick when used with materials that do not have lubricating properties. A cutaway of a plugcock valve is shown in Figure 4 : 12. A special wrench (not shown) is used to turn the shaft.

Butterfly Valves

For large pipes, the butterfly valve is frequently most economical. In this type of valve the opening gate rotates, rather than moving up and down. See Figure 4 : 13.

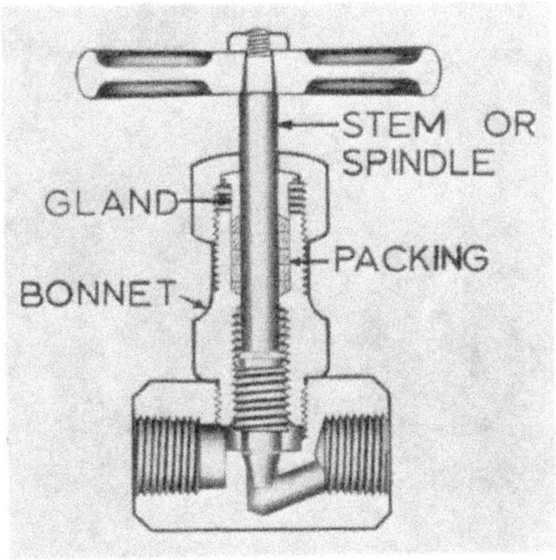

Fig. 4:11. Cutaway of a needle valve (*Crane*).

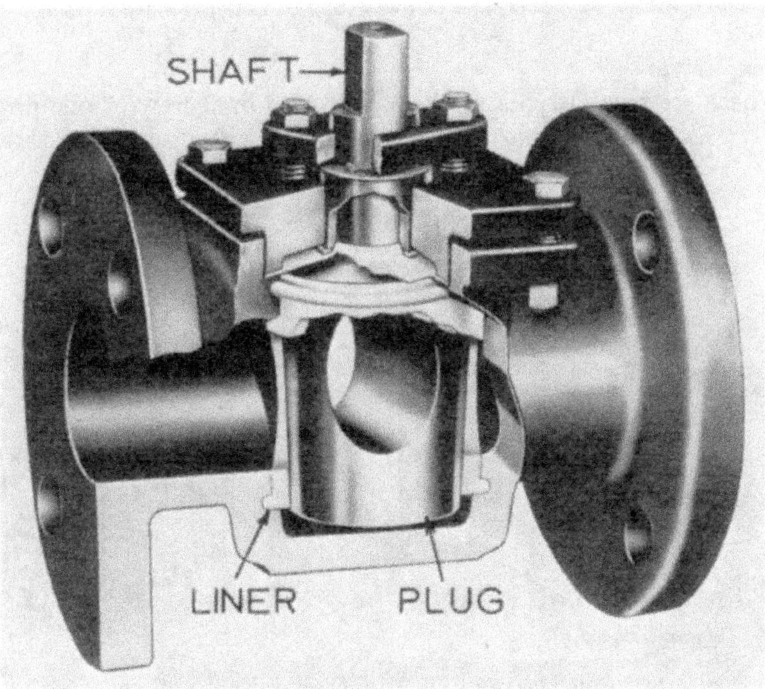

Fig. 4:12. Cutaway of a flanged plugcock valve (*Duriron*).

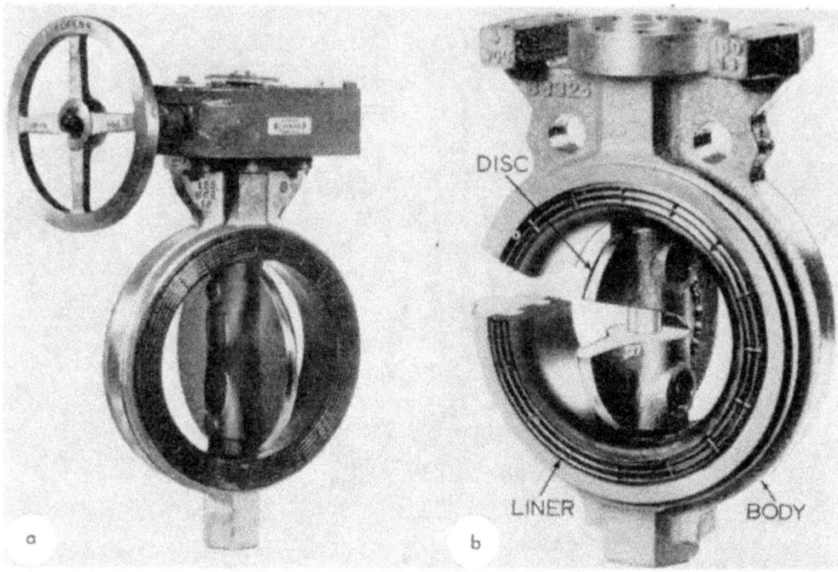

Fig. 4:13. Picture (a) and cutaway (b) of a butterfly valve (*Crane*).

Check Valves

There are several types of check valves to limit flow to one direction. The *swing check* (*a*) and the *lift check* with disc (*b*) are shown

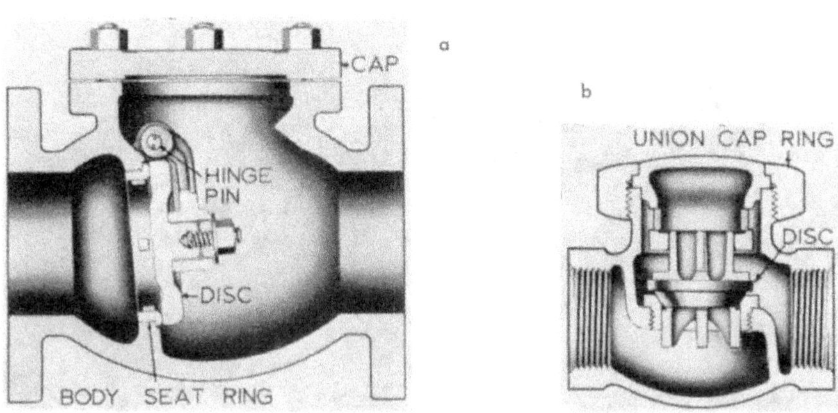

Fig. 4:14. Cutaways of two types of check valve. (a) Swing type; (b) lift type (*Crane*).

in Figure 4 : 14. In place of the disc, some valves use a ball or piston, but the principle is the same. For proper seating, the check valve depends upon the pressure of the line. Against a low drop in pressure or intermittent pressures, they have a tendency to leak.

Diaphragm Valves

There are two basic types of diaphragm valves that protect the bonnet (main section, housing stem of screw and packing) from the liquid being controlled. Both valves thus avoid the need for packing and leakage from a stuffing box.

In the packless diaphragm valve (Figure 4 : 15a), the elastomer of the diaphragm has a hole through its center and is locked between the disc and plunger. In this form, the diaphragm serves only to retain the fluid and provides partial isolation of the bonnet. Closure in the packless valve is effected between the disc and ring-shaped seat.

In the Saunders diaphragm, the bonnet is completely isolated by a continuous diaphragm of elastomer or plastic. The diaphragm is attached to the compressor by means of a stud and closure is effected by compressing the resilient diaphragm against the *weir*.

The general configuration of the Saunders diaphragm valve (Figure 4 : 15b) lends itself to corrosion-resistant lining such as glass, plastic, and other materials more readily than does the packless diaphragm

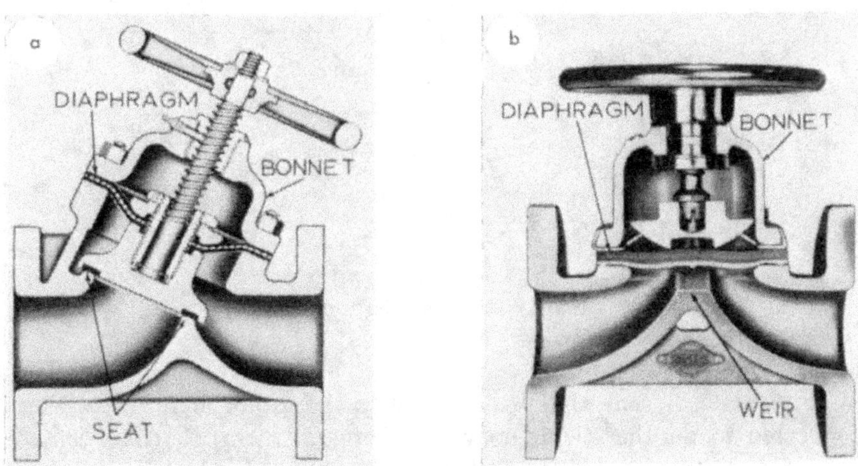

Fig. 4:15. Two types of diaphragm valve. (a) Packless diaphragm valve (*Crane*); (b) Saunders diaphragm valve (*Grinnell*).

valve. Plastic diaphragms, sealed bonnets, indicating stems, and travel stops can be specified on the Saunders diaphragm valve for hazardous applications.

Reasonably-low pressure drops can be expected from both valves.

The valve should never be closed with excessive force; to do so can damage the seating surface.

Flush-Bottom Valves

Where a tank might contain solids and the outlet line must be at the bottom, there is frequently trouble from the solids settling in the short vertical piece of pipe between the tank and the valve. If these solids have a tendency to pack or become hard, a flush-bottom valve can be used. In this type of valve the gate of the valve is flush (even) with the bottom of the tank when the gate is closed. When the valve is opened the gate rises up into the tank. One style of flush-bottom valve is shown in Figure 4 : 16.

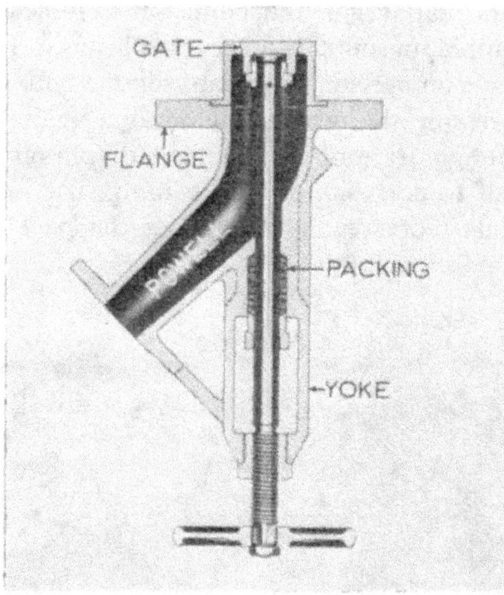

Fig. 4:16. Cutaway of a flush-bottom valve (*Powell*).

Safety Valves

Any vessel or tank that is not open to the atmosphere, or at least connected to another (with no valves between them) that is open to the atmosphere or provided with a safety valve, should have a safety valve. The valve should be sized and installed in accordance with ASA (American Standards Association) or similar piping code. No other valves should be placed between the safety valve and the tank or between the safety valve and the atmosphere.

Two major safety valves are in common use:

1. The *relief valve* is similar to a globe valve except that it has a spring that holds the disc against the seat. If the pressure in the tank reaches a predetermined maximum, the valve will open and allow some gas to escape. As the pressure drops, the valve should close. For minor change of the opening pressure of the relief valve, an adjustment is provided. For any great change in pressure, a different spring may be obtained. It is important that relief valves be checked periodically to make certain that they will open at the predetermined pressure.
2. The *rupture disc* is generally a thin piece of metal held between two flanges. The discs are accurately made to break or rupture within a few percent of the rated pressure. After failure, a new disc must be installed. The discs can be obtained in different metals, carbon, or plastic-coated metal. In general they should be used with vacuum supports. Where large relief lines are required, the rupture disc is often more economical than the relief valve. The rupture disc is also more apt to rupture at the set pressure than is a relief valve, unless the latter is frequently checked. Where sudden large surges of pressure are expected, the rupture disc is superior because it breaks quickly and offers little if any resistance.

STEAM TRAPS

The steam traps are a type of valve. They hold the steam back and let condensate pass. Three different types are in common use:

1. *Thermostatic steam traps* have a valve that expands when steam or hot water is in contact with it and contracts to let the water out when the condensate cools. The trap acts slowly but is quite adequate for light duty.
2. *Inverted bucket traps* are probably the most common. The bucket is held up by steam, thus closing the opening. When the steam condenses, the bucket fills with water and sinks to permit condensate to pass out the opening. The steam trap should be installed with a sediment separator or strainer in the line leading to the trap. A bucket trap, Y-style strainer, and typical piping arrangement are shown in Figures 4 : 17

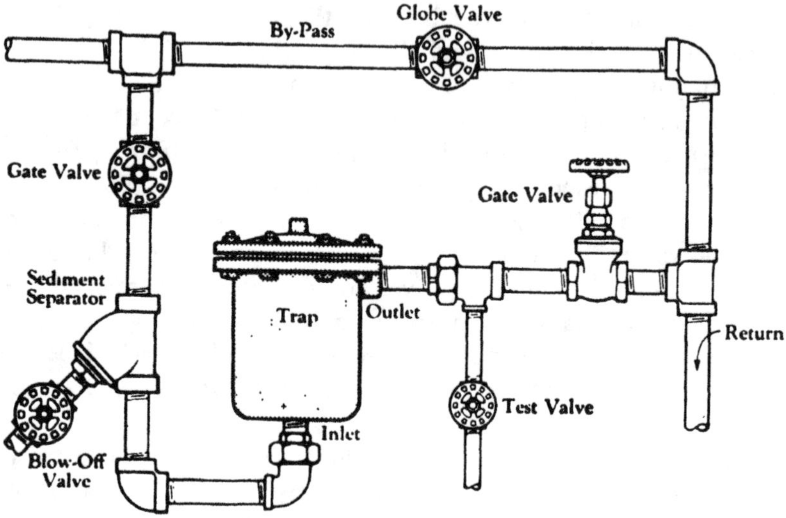

Fig. 4:17. Typical piping arragement for inverted open float steam trap.

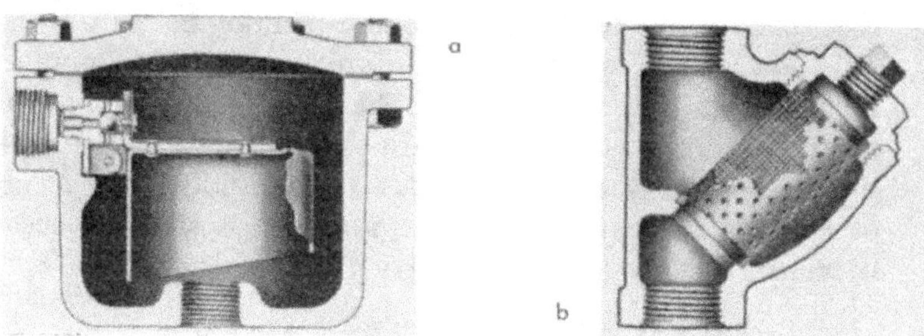

Fig. 4:18. Cutaways of (a) inverted bucket trap in closed position and (b) Y-style strainer (*Crane*).

and 4 : 18. Note that the trap can be removed for repairs while the system is in operation.

3. The *impulse steam trap* (Figure 4 : 19) acts quickly and is thus very good for high-pressure service. This trap is much like a globe valve except that the disc or plug has a hole in it that continuously lets a little condensate pass. If condensate passes, it must come up past the flange, which then lifts the plug and lets the condensate out at the main opening of seat and plug. As the cold condensate leaves and hot condensate and steam enter, the pressure in the control chamber goes up,

permitting the plug to drop.

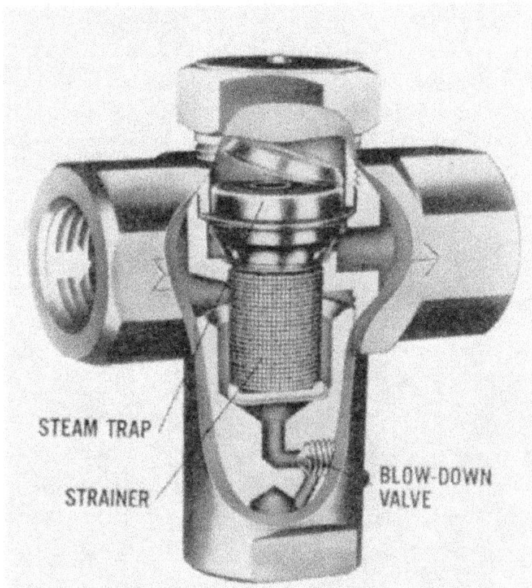

Fig. 4:19. Cutaway of an impulse steam trap (*Yarnall-Waring*).

5. MIXING OF MATERIALS

There are many devices for mixing materials, and the selection of the proper mixer depends upon the reason for mixing. The simplest job is to mix or blend two liquids, which calls for moving the liquids in large volume to all points in a tank. For this purpose, large propellers turning at low speeds might be best. For many other purposes, such as dissolving a solid in a liquid and obtaining uniform turbulence throughout the tank, medium-sized impellers at moderate speeds might be best. There are times when two liquids mix with difficulty and must be emulsified; or when a solid will not dissolve and must be suspended in small particles uniformly distributed in a liquid. For such applications as these, small high-speed impellers are required.

In addition to the choice of impeller or agitator, the size and shape of the tank and the location and shape of heating or cooling surfaces, baffles, etc. must be considered.

BLENDING OF SOLIDS

Ribbon Blender

One type of mixer that is frequently used for blending solids is the *ribbon blender* shown in Figure 5 : 1. This mixer uses helical ribbons to move and mix the material. Solids can be fed to one end of the mixer and continuously removed from the other end.

Twin Shell Blender

Another popular mixer for blending solids is the *twin shell* blender. As shown in Figure 5 : 2, the unit rotates, thereby continually dividing and remixing the batch of solids to be blended. The material to be blended is loaded into the unit through the two hatches at the top of each shell. After mixing, the material is discharged through the valve (not shown) attached to the flange at the bottom of the picture.

Fig. 5:1. Ribbon blender (*Patterson Division, Patterson Industries*).

Fig. 5:2. Twin shell blender (*Patterson-Kelly*).

Paddle Mixers

The paddle mixer is widely used where heavy pastes or doughs are to be mixed. The blades may be at an angle, but are frequently flat. The speed of this mixer is generally limited to 20—150 RPM.

For lighter or less viscous materials, or where the densities of the two materials are appreciably different, stratification will occur with this type of mixer. Where the blades are short, as in Figure 5 : 3, it is often called a *pug mill.*

Fig. 5:3. Paddle mixer (*Patterson Division, Patterson Industries*).

MIXING OF FLUIDS

Flow Mixers

If the job is a very easy one, such as mixing natural gas and air or gasoline and air, the two might simply be brought together in a piece of pipe. The pipe could also have baffles, or the materials could enter through an injector mixer.

If two liquids are introduced to the suction of a centrifugal pump, the turbulence of such a pump can effect very good mixing.

In-line-mixers that consist of a large section of pipe with a propeller and baffles can be purchased (Figure 5 : 4). All of these are fine when

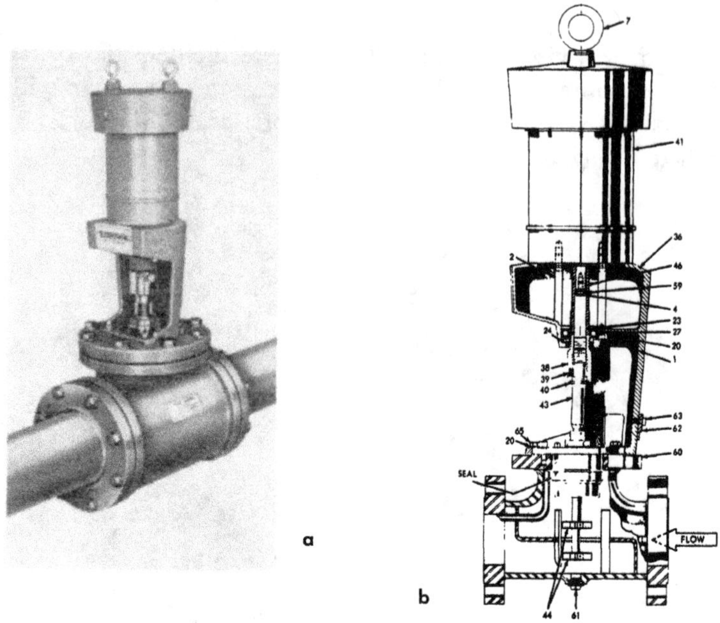

Fig. 5:4. Cutaway (a) and picture (b) of an in-line mixer. (1), (2), (65) Hex head cap screw; (4) shaft screw and lock washer; (7) motor eye bolt; (20) washer; (23) retaining ring; (24) oil seal; (27) ball bearing; (36) housing; (38) drive shaft; (39) set screw; (40) thrust ring; (41) impeller assembly; (46) motor key; (59) washer; (60) vessel assembly; (61) drain plug; (62) shaft support plate 2; (63) socket head cap screw.

the blending or mixing is to be almost instantaneous but if any length of time is required for proper mixing (called *holding* or *residence time*), a tank is required.

One method in common use is to allow the material that is to be mixed to flow continuously out of the tank through a pump and back into it. If a heat exchanger for heating or cooling the material can be inserted in this recycling loop, an additional operation can be achieved.

Propeller Mixers

The propeller mixer is in common use with many variations. For large tanks, the propeller mixer can be mounted at the side near the bottom, and keep a gentle swirling motion in tanks of about 100,000 gallons. The mixer can be mounted on top of a smaller tank and, with one or more propellers and proper baffling, keep good agitation and turbulence going throughout the tank. The propellers can also be surrounded with a tube, called a *draft tube* (Figure 5 : 5), which will aid in preventing stratification. Generally, the use of a propeller is limited to light or moderately viscous liquids where intimate mixing is desired.

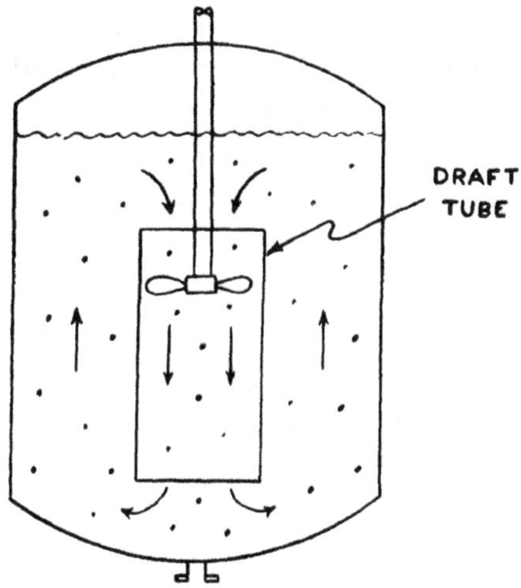

Fig. 5:5. Cutaway of a draft tube system.

Turbine Mixers

The turbine mixer, or *centrifugal-impeller mixer*, is one of the most versatile types. When submerged, it acts much like a centrifugal pump. The impeller is usually small and operated at relatively high speeds. The impeller may have straight, pitched, or curved blades and may be of the *open*, *closed* (shrouded) or *semi-open* type. Open impellers have a higher shear. The semi-open is excellent when used with a gas inlet pipe at the eye of the impeller for adding gases to a liquid.

The turbine mixer is very good for viscous liquids, for rapid dissolving, for suspending heavy slurries, and for mixing in tanks or containers of irregular shape.

Shaft Seals

Essentially the same problems are encountered with shaft seals for mixing equipment as with pump shaft seals. The discussion of shaft seals in Section 2 of this manual also applies to this section.

OPERATING NOTES

Although it is sometimes safe to operate an agitator without liquid in the tank, it is usually not without risk. No difficulties are experienced with low-speed, vertically mounted mixers, but with higher-speed mixers (and especially with agitators mounted at an angle or horizontally) deflection of the shaft can start a whipping motion and serious damage to the shaft can result.

The system of high-pressure seals must be functioning and up to pressure before the pressure in the tank is raised, particularly on high-pressure reactors, where double mechanical seals are used. A shaft seal is generally not tight unless it is rotating, and mechanical seals in particular will leak when not turning.

6. HEATING AND COOLING

GENERAL

With the possible exception of pumps, heat exchangers are the most-widely used form of major equipment in chemical plants. Although the design and type of the heat exchanger may vary, the principle is the same for all—to transfer a quantity of heat from one material to another. Three important factors are involved:

1. The first factor is resistance to the transfer of heat. Let us consider cold water on one side of a metal wall and a hot liquid on the other side.

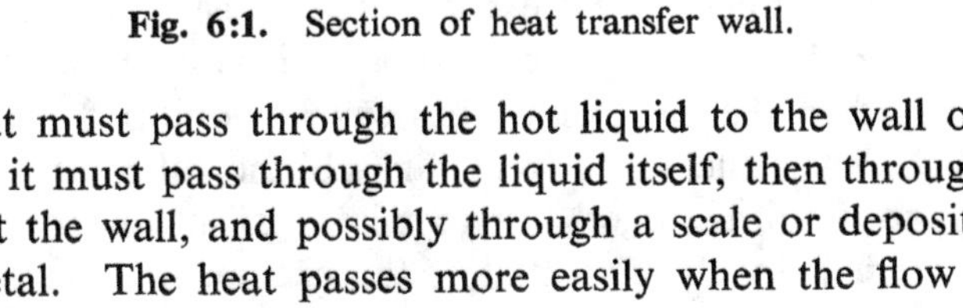

Fig. 6:1. Section of heat transfer wall.

The heat must pass through the hot liquid to the wall of metal. To do this it must pass through the liquid itself, then through a film of liquid at the wall, and possibly through a scale or deposit on the wall of metal. The heat passes more easily when the flow of fluid is sufficiently fast to cause turbulence or eddys than with a slow flow, which would give a smooth or laminar movement. The heat must then pass through the wall itself.

Everyone is familiar with the fact that the handle of a silver spoon will become hotter than the handle of a stainless steel spoon in the same cup of hot coffee; and also with the fact that the farther a poker is grasped from its red-hot end, the cooler is the metal. These same principles apply to a heat exchanger. The thicker the wall, the greater the drop in temperature through the wall. The drop in temperature will also depend upon what metal is used. The heat must then pass across any scale or deposit on the other side of the wall, through another film of water, and finally into the water itself. A drop in temperature is therefore created by this resistance. In use, a heat exchanger frequently transfers less and less heat as scale or deposists build up on the wall. This is called *fouling*; when it becomes excessive, the heat exchanger must be cleaned.

2. The second factor is area of transfer surface available. A very fine wire can be held in the hand, even if the red hot end of the wire is only a foot away; this cannot be done with a heavy rod of the same material.

3. The third factor is the overall difference between the temperature of the cold water and temperature of the hot liquid. The greater the difference in temperature, the faster the heat is transferred. Heat transfer may be expressed by the formula

$$Q = UA\Delta T$$

where Q is the quantity of heat transferred and is expressed in BTU/hr; U is the resistance or lack of resistance. Actually, it decreases as resistance increases. U is generally called the *coefficient of heat transfer;* A is the surface area of the heat exchanger tubes; and ΔT is the over-all difference in temperature expressed in °F. A normal value of U would be in the range 50 to 150 BTU/sq. ft.-°F.

HEAT EXCHANGERS

Open or Atmospheric Exchangers

Open, or atmospheric, heat exchangers consist of tubes that carry the material to be heated or cooled, and in which the material is simply exposed to the air. The tubes can be equipped with fins and the air forced over the tubes by a fan. If the tubes are formed by stamping and welding together two plates of metal so that many tubes are formed by the two plates, the exchanger is called a plate cooler.

This type of exchanger is limited to cooling very hot materials, or to heating very cold materials such as liquid nitrogen. It is also limited to uses where the final temperature of the material being heated or cooled does not require accurate control. As the temperature of the air might vary from about 100°F in the summer to below 0°F in the winter, the performance of the heat exchanger will vary considerably.

Jackets

Frequently, one tank will be surrounded by another, forming a space between the two into which a heating or cooling medium can be placed. In general, the inner tank of the jacket does not furnish many square feet of heating surface. Depending upon its shape, the jacket of a 100-gallon tank will furnish about 30 square feet, or 0.3 square feet to the gallon; on the same basis, a 1,000 gallon tank will furnish about 150 square feet, or 0.15 square feet to the gallon. On larger tanks, therefore, the jacket becomes less and less practical. The jacket is also expensive compared with other types of heat exchanger.

Internal Coils

A coil inside a tank is more economical than a jacket, especially on larger tanks. The coil is more effective if the tank is eqipped with an agitator designed in conjunction with the coil. Where the temperature must be very exactly controlled—that is, where hot spots cannot be tolerated—the coil may not be satisfactory. It is also much more difficult to clean or repair a coil inside a tank.

Reboilers

As the name implies, the *reboiler* is a heat exchanger designed to reboil or reheat a material. It is used near the bottom of a fractionating column and frequently at the bottom of a *stripper* or tank. There are a number of types of reboiler:

1. An *internal coil reboiler* is one of the most economical, but the number of square feet of coil is limited. Hot spots may be present, and if the coils are close together any solids present may gradually plug the spaces between the coils.
2. An *external reboiler* may be used where the liquid flows out of the bottom of the column into the bottom of a heat exchanger and the vapors rise and enter the column. This type is generally called a *kettle* (Figure 6 : 2a). If the external heat exchanger is so connected that the vapors and heated liquid act to provide circulation, and both return to the column, the reboiler is called a thermosyphon (Figure 6 : 2b).

A pump is added to the line between the bottom of the column and the reboiler, particularly where solids might be present, where the quantity of heat required is high, or where hot spots must be avoided. This ensures a high rate of flow, which will keep solids in suspension, reduce hot spots in the reboiler, and increases overall

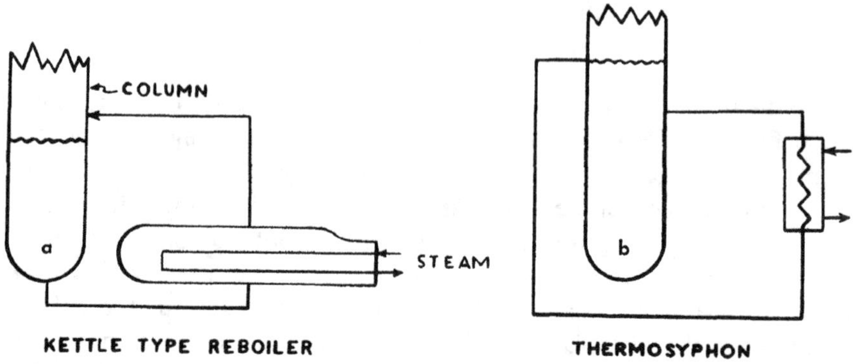

Fig. 6:2. Common types of reboiler. (a) Kettle type; (b) thermosyphon.

heating efficiency.

Condensers

A heat exchanger used to cool gases for the purpose of liquefying or condensing one or more of the gases is often referred to as a condenser. The shell-and-tube type of heat exchanger discussed later in this chapter is generally used for this purpose. The condenser may be mounted vertically or horizontally but the water must be in the tubes and the gases in the shell. One prime reason for this is that while the water will gradually foul the tubes, it is much easier to clean the inside of the tubes than the outside. It is important that the water enters at the bottom so that the tubes are full of water. If the water were allowed to vaporize, the solids dissolved in the water would be left behind as a scale.

Double-Pipe Exchangers

An arrangement of one pipe within another, where the material to be heated or cooled is in the inner pipe and the other material is in the annular space between the two pipes, makes a very inexpensive exchanger. If the inner pipe has fins out into the annular space, the efficiency is greatly improved. The double pipe exchanger is low in cost, but the pressure drop is rather high.

Shell-And-Tube Exchangers

The shell-and-tube exchangers, which are extensively used, can vary considerably. One of the least complicated types, a single-pass with fixed tube sheets, is shown in Figure 6 : 3.

In the model shown, the cold material to be heated might flow into the exchanger through nozzle A, through the tubes, and out of nozzle B. The hot material flows into the exchanger through nozzle D, is forced around the tubes by the baffles, and leaves by nozzle C.

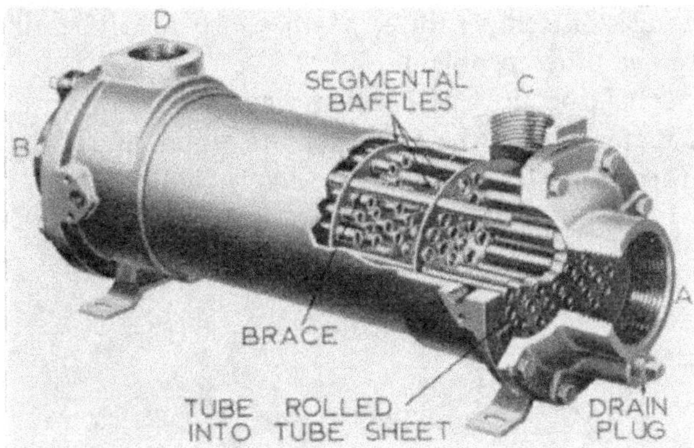

Fig. 6:3. Cutaway of a single pass, fixed tube sheet shell-and-tube exchanger (*Industrial Division, American Standard*).

If the difference in temperature between the two materials is very high, a rather large stress might be placed on the tube sheets by the difference in expansion of the tubes and the shell. In such circumstances, a four-pass floating-head exchanger might more profitably be used.

If one fluid would foul the surface more rapidly than the other, it is usually run through the tubes since it is easier to clean the tubes than the shell. If one fluid is very corrosive, it is also run through the tubes, since in this case only the tubes would need to be made of the more expensive, corrosion-resistant material.

In many models, liquid flows in at the bottom nozzle and out at the top nozzle to avoid trapping air; a condensing gas flows in the opposite direction.

Interchangers

Frequently, the feed to a system requires heating while the stream out requires cooling. A heat exchanger can permit heat to be exchanged between the two, thus saving on both steam and cooling water. These exchangers are generally called *interchangers*. Either the double pipe or shell-and-tube type exchanger may be used for this purpose.

SUGGESTIONS FOR OPERATION

Where possible, start the flow of cool material first and cut off the flow of cool material last. This will prevent overheating, which could contribute to the fouling of tubes.

Watch changes in temperature, as they are an indication of low flows, fouling, or other problems.

When reducing the flow of cooling water in order to obtain less cooling of a material, use the valve on the water line out of the exchanger for throttling, unless otherwise instructed.

7. SEPARATION OF MATERIALS

SOLIDS FROM SOLIDS

There are several common methods that may be used to separate one solid from another.

1. Solution—in which a solvent dissolves one substance and leaves the other.
2. Melting—in which one substance melts so that the other can be filtered out.
3. Vaporization—in which one substance vaporizes (sublime) and leaves the other.
4. Flotation—in which a liquid is be used to float the lighter solid away from the heavier one.

A study of the chemical and physical properties of the solids is necessary to determine which of the preceding methods is best.

Frequently in the chemical plant it is necessary to separate a product according to size. To remove dust or fine particles from a solid, the cyclone separator (see *Air Conveyors* in Section 1) can be used.

Rotary Screen

One of the most commonly used pieces of equipment for separation according to size is the rotary screen separator (Figure 7 : 1). With this equipment, the solid is fed into a box, which is agitated by a rotary movement. The box may have almost any number of screens so as to separate the product into lots containing particles of a certain size range. Usually, plastic balls are used to keep the screens from becoming blinded or plugged.

SOLIDS FROM LIQUIDS

Plate-and-Frame Filter

One of the oldest methods of separating solids from liquids is the plate-and-frame filter, shown in Figure 7 : 2.

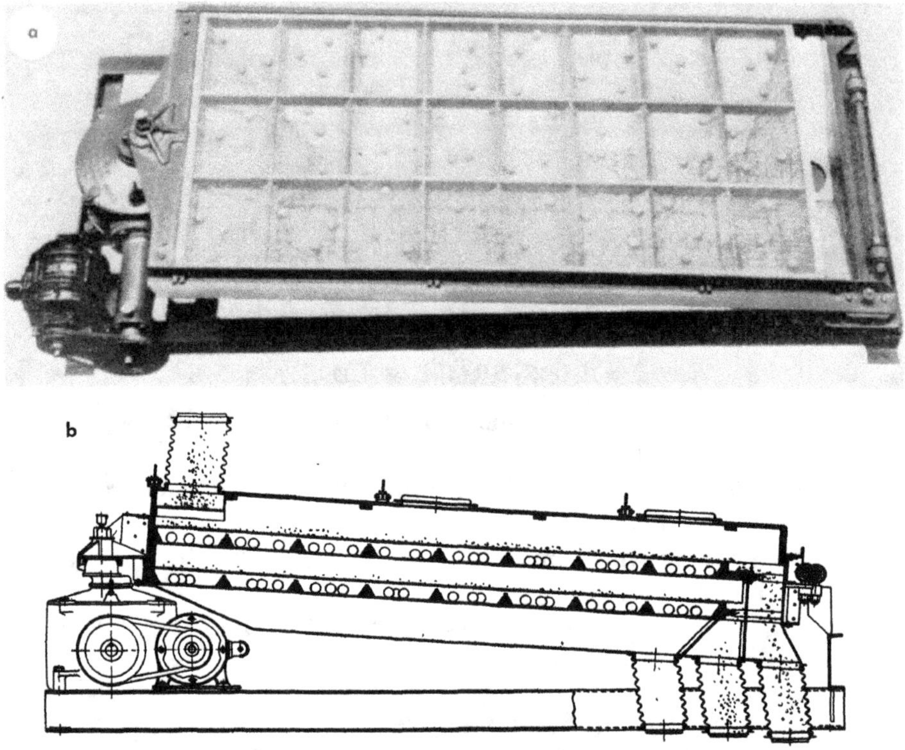

Fig. 7:1. Rotary screen separator, (a) Top view; (b) cross section (*Rotex*: *Orville Simpson*).

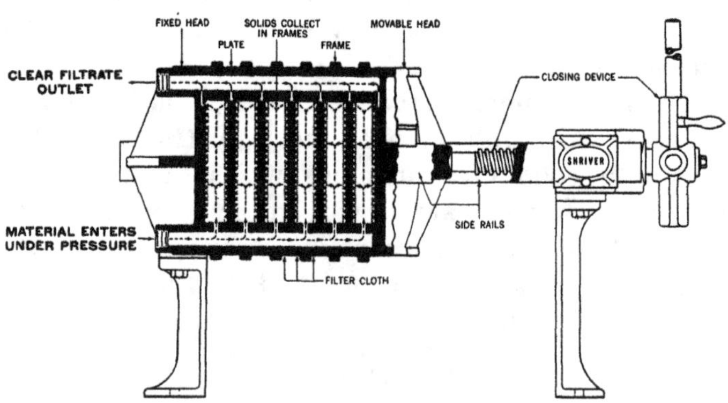

Fig. 7:2. Cutaway of a plate-and-frame filter (*T. Shriver*).

This is a batch unit, generally limited to small volumes of solids. Washing of the solids is not too efficient, and generally the solids retain a rather high percentage of liquid. Some of the newer units are mechanized to provide faster removal of the cake.

Batch Centrifugal Filter

The batch centrifugal filter has many variations in design. Generally, the basket is 12-48 inches in diameter, although other sizes, from 6-84 inches in diameter, are not uncommon. The motor may be above the basket, in which case the filter is known as a *suspended basket* centrifugal. With the motor below the basket it is generally called *under-driven.* The suspended basket machine is more flexible as it can be more easily automated and equipped with bottom discharge. The under-driven basket is, however, easier to change, clean, and make vapor-tight, and takes less head room. Horizontal basket units are also available.

A 30-inch basket, for example, is generally direct-driven by a motor of about 25 hp at 1200 RPM, and is equipped with a brake for fast deceleration. Frequently a multispeed motor or variable drive is provided.

The basket can be a solid bowl into which the liquid is forced up the wall and over the rim, or a perforated bowl in which the liquid is forced through the cake and holes in the basket.

In operation, the centrifugal filter is run at moderate speed, and the slurry to be filtered is fed at such a rate that liquid is apparent up to almost the full side of the perforated basket. If fed too slowly, the cake will be very thick near the bottom and will be difficult to wash. If fed too fast, the solids will be carried over the top of the basket. When the cake is sufficiently thick, the feeding is stopped and the cake is allowed to spin in order to remove as much of the liquid as possible. Often a higher speed is used at this point. The cake is then washed and allowed to spin to remove the wash, the basket is stopped, and the cake is dug out. Frequently a blade or plow is provided to loosen the cake for removal and the basket rotated at a very low speed during this operation.

The centrifugal filter does not do a really good job of washing the cake, but it does provide a reasonably dry cake compared with many other types.

Continuous Centrifugal Filters

There are two newer styles of centrifugal filter in common use. One is the centrifugal decanter, which is a horizontal bowl machine, and the other is the conical centrifuge, which has a perforated, vertical, conical-shaped bowl. Both of these machines have spiral scrapers for continuous removal of the cake. Although these two machines have good capacities, they both have the problem of considerable agitation, which tends to keep small particles in suspension, and thus a clear filtrate is not always possible.

Figure 7 : 3 shows details of the centrifugal decanter.

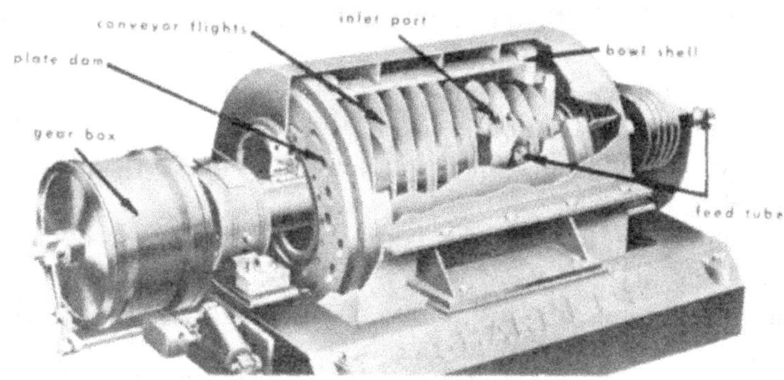

Fig. 7:3. Cutaway of a centrifugal decanter (*Sharples Division, Pennsalt Chemicals*).

The feed enters the hollow shaft and in the "pond" section separates the solids and liquid. The liquid flows up and over the filtrate ports, while the spiral scraper moves the solids to the washing section and then to the "beach," where they drain. There are adjustments to increase or decrease the length of the pond compared with the length of the beach, and the speed of the scroll or scraper can be varied relative to the speed of the solid bowl. Washing in this unit is not generally very effective.

A vertical centrifugal decanter that may be preferred for operation under moderate pressures is also available. The vertical, conical-shaped centrifugal is similar to the decanter, except that is is a perforated basket. It is generally called a *continuous conical* or *cone* centrifugal filter.

Centrifuges

Although centrifugal separators are often called *centrifuges*, the real centrifuge is generally a high-speed machine of small diameter. It is used to separate liquids of different densities, to break emulsions, and to remove fine solids in suspension. A laboratory unit, consisting of a solid tubular bowl 3-6 inches in diameter and about five feet long, may operate at 15,000 RPM and develop a centrifugal force of 13,000 times gravity. Units are available up to 100,000 times gravity. They can be used for 50-500 gallons an hour, but solids must be removed one batch at a time; therefore, except in operations where essentially no solids will collect, they are only semi-continuous.

Disc-bowl centrifuges are larger but these do not develop the centrifugal force of the small-diameter units mentioned. Figure 7 : 4 shows a continuous type of centrifuge.

The feed enters at the top and is accelerated by vanes as it passes down. The liquid travels up the discs, while the solids travel down. The solids, in the form of a wet sludge, are generally recycled to the unit, and dryness or percentage of solids in the sludge is controlled in this manner. As large particles will plug the nozzles, the solids must be quite wet to flow.

Frequently another type of unit is used to remove large particles before they enter the disc-bowl or continuous centrifuge, and the concentrated solids from this unit are then sent to a rotary vacuum filter. The purpose of the disc-bowl centrifuge, therefore, is more

a

b

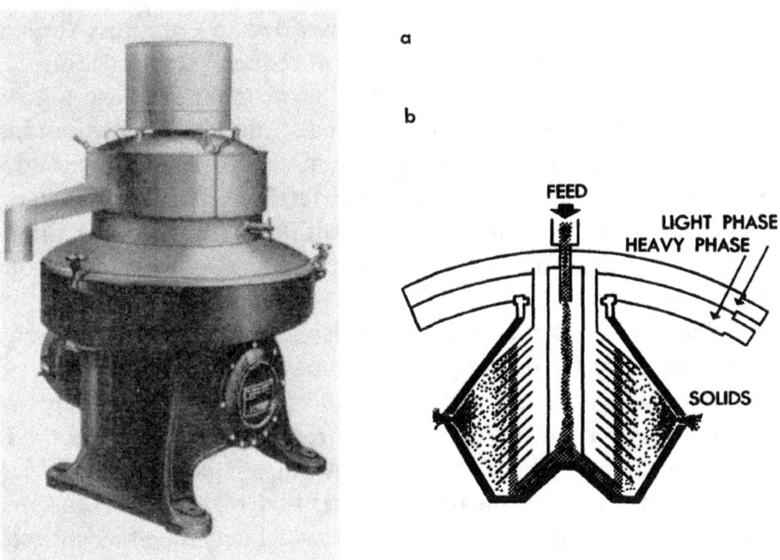

Fig. 7:4. Picture (a) and cutaway (b) of a disc-bowl centrifuge (*Sharples Division, Pennsalt Chemicals*).

a concentrator of solids that settle with difficulty because of their fineness and because their specific gravity is close to that of the liquid.

Rotary Vacuum Filter

There are several variations of the rotary vacuum filter, such as *blowback*, *precoat*, and *endless belt*.

1. A *blowback* filter is shown in diagrammatic form in Figure 7 : 5. The slurry is fed continuously to maintain a certain level, and generally a slow-moving rake-type of agitator keeps the solids from settling out. The drum rotates slowly and a vacuum applied at the nozzle V_1 sucks the liquid through the submerged segments numbered 1-5. The cake gradually builds up on the drum. The speed of rotation of the drum can be varied to regulate the thickness of the cake. The cake can be washed as it moves past the wash nozzles. A vacuum applied at nozzle V_2 sucks the wash water through the cake on segments numbered 6-11.

By use of two separate vacuum points, the filtrate or mother liquor can be collected separately from the wash water. The two vacuum points are isolated from each other by the bridges or blocks numbered B1, B2, and B3 in the rotary valve. The segments 9-11 suck the cake as dry as possible and in this dewatering section cracks in the cake can occur that would allow air to pass and prevent proper drying. A *compression roll* is sometimes provided to compress the cake to a uniform mass, which eliminates or reduces the cracking. At nozzle V_3 an air pressure of about 5 *psig* blows the cake loose. A doctor knife or scraper helps to remove the cake and directs it to a chute.

If the cake tends to pack too tightly, filter aid is sometimes added to the slurry before it is pumped to the filter. This filter aid can be any solid material that will not react with the liquid being filtered, that will not pack, and can easily be drained. A filter aid is often called an *admix*.

2. Precoat filter can be used if the solids are slimy or otherwise impossible to filter in the regular vacuum filter. A filter aid, such as calcium sulfate, is mixed with water and pumped to the filter. The cake is allowed to build up to two or three inches in thickness. The slurry is then filtered, but instead of blowing off the cake, the doctor knife slowly advances by means of gears, and on each revolution of the drum cuts off a thin sliver of the cake on which slime has collected.

3. The *endless belt* variation does not require a precoat. The drum is covered with an endless belt of suitable fabric, which, after the cake is picked up, washed, and dewatered, passes over a roll.

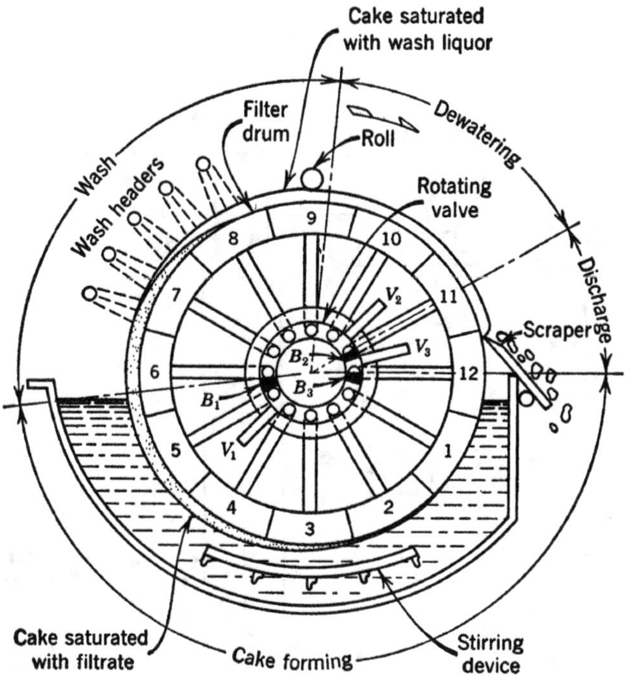

Fig. 7:5. Diagram of a Blowback Rotary Vacuum Filter (*Eimco*).

This loosens the cake and the belt can then be washed before returning to the vat.

The prime use of this type of filter is with materials that should not be contaminated with filter aid, but where the cake tends to "blind" (plug) the filter cloth. Strings can be substituted for the belt to aid in lifting off the cake, in which case the filter is known as a *string filter*.

The rotary vacuum filters provide excellent washing of the cake and can economically handle very large volumes of cake. The resulting cake, however, is generally much wetter than cake from a centrifugal unit. A process flowsheet showing the general arrangement of a rotary vacuum filter is shown in Figure 7 : 6. Where it is not necessary to wash the cake or segregate the wash water from the mother liquor, the rotary valve and extra receiver system can be eliminated.

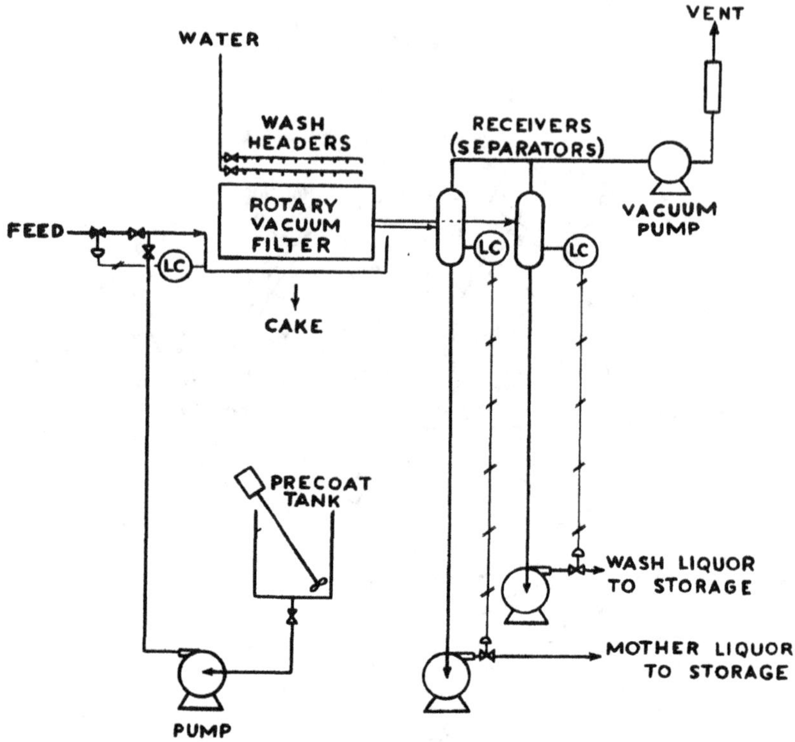

Fig. 7:6. Flowsheet of a typical rotary vacuum filter system.

LIQUIDS FROM LIQUIDS

Non-Miscible Liquids

Liquids that do not mix, such as oil and water, can be separated in tanks by allowing the lighter one to float to the top. If the specific gravities of the two are close together or a large volume must be separated continuously, a vessel packed with glass fiber or similar material, called a *coalescer*, can be used.

Miscible Liquids

Where one liquid material dissolves in another, several methods of separation can be used.

1. *Distillation* is the most common way of separating one liquid from another. This is the process of boiling off the gas or vapor from a liquid and then condensing this into a liquid again by cooling. It is an excellent way to separate a liquid from impurities or to obtain

it in its pure state.

2. *Extraction:* In a *liquid-liquid extraction* the material dissolved in the liquid is transferred to a second liquid in which it is soluble. To be effective, this second liquid must be insoluble, or only slightly soluble, in the first liquid.

3. *Adsorption:* Sometimes a separation can be made by *adsorption.* For example, small amounts of water dissolved in benzene can be removed by pumping the liquid through a column of *molecular sieve.* This consists of a zeolite, chemically and physically designed to hold onto the small molecules of water and allow the large benzene molecules to pass. After the molecular sieve is saturated with water it can be heated to drive the water off and can then be used again. Water in the benzene can be reduced very easily in this manner to less than five parts per million.

Fractional Distillation

As we said before, distillation is the separation of a liquid from impurities by heating and vaporizing. If we take plain (natural) water and heat it to its boiling point we can collect pure (distilled) water by condensing and collecting the vapors.

If we heat a mixture of methanol (methyl alcohol) and water, however, both the water and the methanol will vaporize. Assuming that we started heating a mixture of 90 percent water and 10 percent methanol, the first part (20 percent) vaporized would consist of 60 percent water and 40 percent methanol, and the liquid left behind would be 98 percent water and 2 percent methanol. The reason for this is that the methanol is more volatile. At the temperature at which we started to boil the mixture, the vapor pressure of the methanol was much higher than the vapor pressure of the water, so that most of the methanol came off, while only a small part of the water came off. The boiling point of pure methanol is about 148°F (55°C); the boiling point of water is 212°F (100°C); the mixture of 10 percent methanol and 90 percent water boils at 198°F (92°C). If we continued to heat the remaining liquid we would find that the temperature at which the liquid boils would increase until it reached about 212°F. Naturally, all of the methanol and water would finally be vaporized.

If we collected only the first 20 percent that was vaporized and condensed (60 percent water and 40 percent methanol) and heated this liquid, we would find it boiling around 175°F (80°C); and the first 45 percent vaporized would consist of 22 percent water and 78 percent methanol. If we again collected this small part and heated it, we would find it boiling around 158°F (70°C), and the first 25 percent vaporized would consist of 90 percent methanol and 10 percent water. We could keep this up until pure methanol was obtained.

The series of distillations described is called *rectification* or *fractional distillation.* This fractional distillation could be done in a column that has trays so designed that the vapor coming up is forced by a bubble cap to bubble through the liquid. This condenses the vapor, the heat from which will create new vapor that will be richer in alcohol than the vapor that came up. With a sufficient number of trays, pure methanol could be taken off at the top of the column and water free of methanol at the bottom (Figure 7 : 7). The feed to the column would be on the tray where the composition of the original liquid is about the same as that of the liquid on the tray. As the trays below the feed tray remove the lighter, or more volatile, component, they are called the *stripping section.* The trays above the feed tray are called the *rectifying section.*

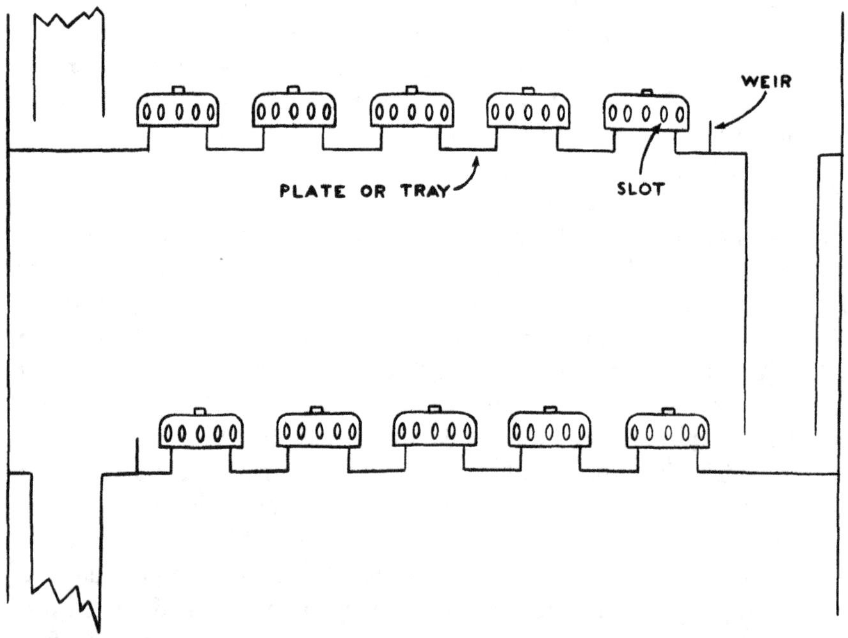

Fig. 7:7. Cross section of a distillation column.

Refluxing

The performance of the column depends upon the contact between the vapors coming up and the liquid coming down. There must be sufficient of both liquid and vapor to ensure good mixing. If all the vapor coming off at the top of the column could be condensed and retained in a tank, and if the column were well insulated so that no vapors were condensing, there would be no liquid flowing down at the top of the column. We therefore condense the vapors coming off at the top of the column and return part of this liquid to the top of the column. This is called *refluxing*. The *reflux ratio* of a column is the amount of liquid returned to the top of the column (L, expressed as *g/min* or *lb/hr*) divided by the amount of condensed liquid removed from the column (D, same units).

The more refluxed fluid that returns into the top of the column, the more efficiently each tray works. This means that we need fewer trays in the column. To put a column in what might be called "neutral gear" or "standby," the feed can be stopped and the column put on *total reflux*. This means that all the vapor from the top of the column would be condensed and put back into the top of the column. If the

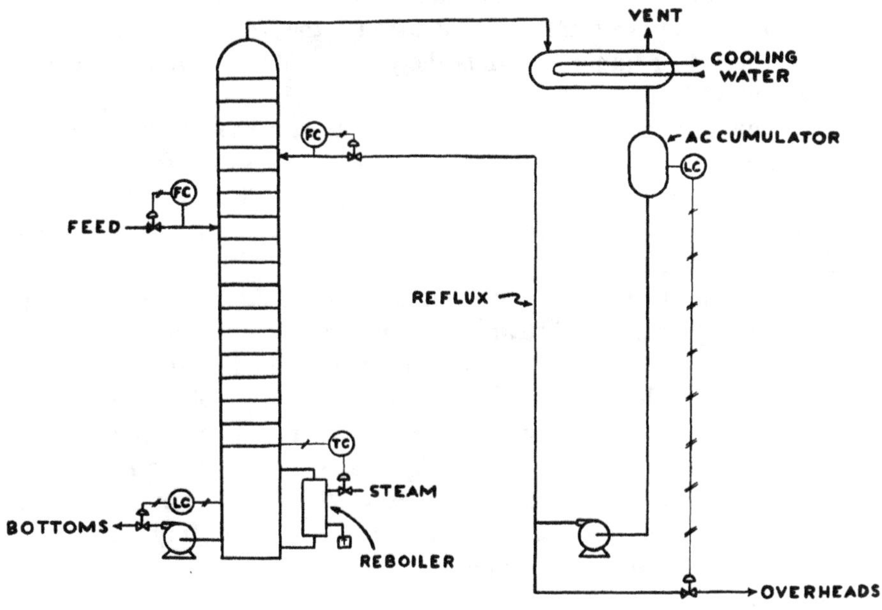

Fig. 7:8. Flowsheet of a typical distillation column.

column is on total reflux, there is no production. As the quantity of reflux is decreased, or the reflux ratio is lowered, production increases, use of steam and cooling water decrease, but less separation is obtained. There is therefore a minimum to which the reflux ratio can be reduced and still obtain the desired separation. Figure 7 : 8 shows a typical arrangement of a distillation column.

Types Of Distillation Tray

The trays used in distillation columns are of several types:

1. *Bubble Cap Trays* were the first type developed. Figure 7 : 9d shows a small complete tray. Referring to this photograph and the cross section drawing in Figure 7 : 7, it is obvious that the gas or vapor rising in the column must be sufficient to prevent the liquid from running down the slots in the bubble caps, or there will not be a sufficient level of liquid on the trays for good contact with the vapor—and a pressure drop will be created across each tray. In a column with a large number of trays, the pressure, and therefore temperature, at the bottom of the column will much higher than at the top. This is not always desirable.

2. *Sieve Trays* were designed to reduce the pressure drop across each tray. They consist of trays similar to those shown in Figure 7 : 7 except that the bubble caps are replaced with many small holes in the plate. These small holes distribute the vapor coming up so that the height of liquid and thus the pressure drop) across each tray can be reduced.

3. *Valve Trays* were perfected in order to vary the opening through which the vapor passed. In the bubble cap and sieve tray a certain quantity of vapor was necessary. As previously discussed, the vapor is generated by steam in the reboiler and vapor must be condensed with cooling water.

If a low production rate were desired, it would still be necessary to use large quantities of steam and cooling water in order to maintain the necessary quantity of vapor to prevent the liquid from draining through the slots or holes and lowering the level of liquid. With the valve trays, the opening is changed according to the quantity of vapor available. At high production rates, the large quantity of vapor lifts the plate shown in Figures 7 : 9c and d. At low rates, the plate drops to reduce the opening. Complete tray assembles are shown in Figures 7 : 9a and b.

Packed Columns

If many trays are required, the types just discussed are very good,

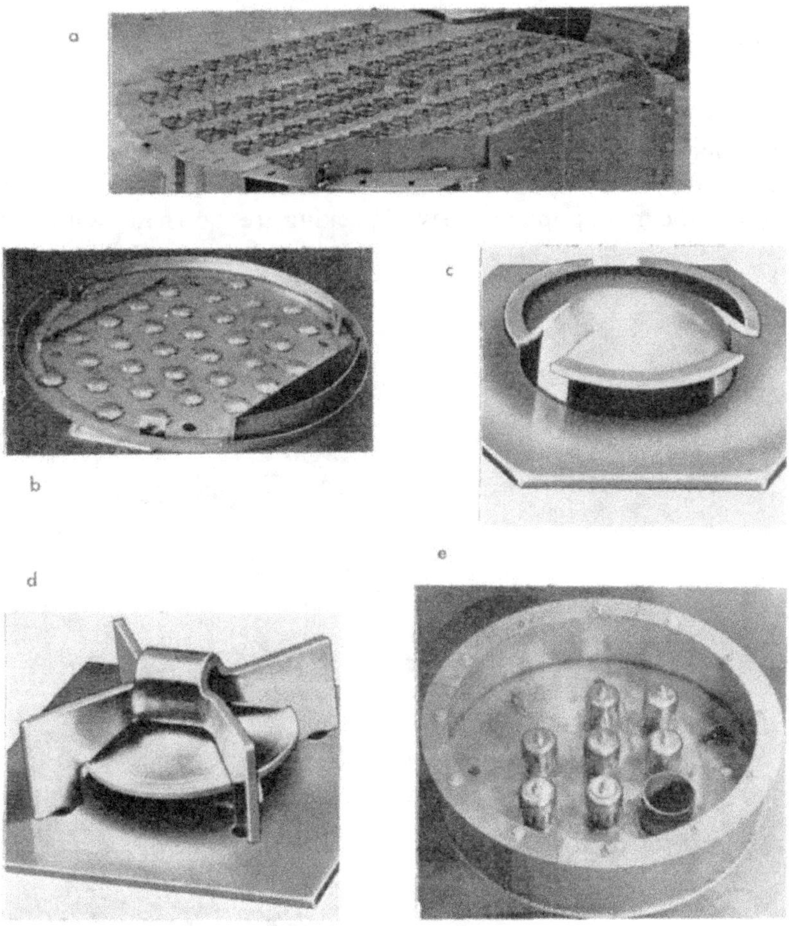

Fig. 7:9. Types of distillation tray. (a), (b) Complete valve type trays; (c), (d) individual valve type caps; (e) complete bubble cap type tray (*Koch Engineering*).

but if only five to ten trays are required, packing can be used in place of them. Trays might be about 12 to 15 inches apart. Where packing is used, ceramic or metal shapes can be stacked or dumped into a column and, depending upon which shape is used and how it is placed, each foot or two of packing will have the same effect as a tray. As the drop in pressure across packing is generally much higher than across a tray, we usually use packing only where the equivalent of several trays is required.

One problem with the packing is that depending upon the type of packing and the diameter of the distillation column a certain quantity of liquid and gas is required for satisfactory operation. These flow rates are more critical than in an equivalent column of trays. Where the liquid and gas flow rates would require a column of less than 24 inches, the packing is superior because it is difficult to make and place trays in a column less than 24 inches in diameter. Several of the more popular types of packing are shown in Figure 7:10.

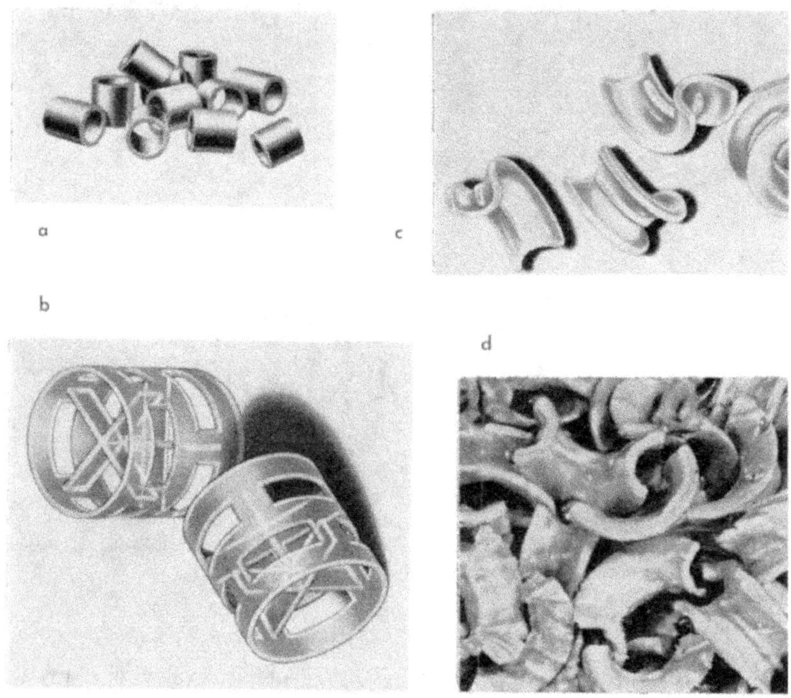

Fig. 7:10. Various types of packing for distillation columns. (a) Raschig rings; (*b*) pall rings; (*c*) intalox saddles; (*d*) dumped intalox saddles under irrigated conditions (*U. S. Stoneware*).

Strippers

Occasionally small quantities of a volatile material must be removed from an essentially non-volatile liquid. Letting this material flow down through a packed tower, generally counter-current to a rising stream of gas (such as warm air), is an economical means of removing the volatile component. The tower is generally referred to as a *stripper* (not to be confused with the stripping section of a fractionation tower previously discussed).

GASES FROM GASES

Absorption

One of the least expensive methods of separating one gas from another is by absorbing one of the gases in a liqud or on a carbon bed. Naturally, the other gas must be absorbed either only slightly or not at all in the given liquid or carbon bed.

For example, in the process of making ammonia from natural gas, carbon dioxide must be removed from a mixture of hydrogen and nitrogen. The gases flow up through a column with the liquid monoethanolamine (MEA) flowing down through the column. The MEA reacts with, and removes, the carbon dioxide. The MEA can then be heated to drive off the carbon dioxide, cooled, and used again.

Cooling

A rather common method of separating gases is by cooling. As most gases have dew points that differ considerably, this method has wide application although it is a relatively expensive process.

For example, in liquid oxygen plants, air is cooled to separate the various gases by condensing them. After the nitrogen condenses around −346°F (−210°C), the remaining gases can be further cooled and oxygen will condense around −360°F (−218°C). On further cooling, the neon would condense at around −316°F (−248°C).

Molecular Sieve

One of the newer methods is the use of molecular sieve, which will absorb gas of small molecules such as carbon dioxide and allow the larger molecules of gas to pass through. The gases are allowed to flow upward through a column containing the beads of molecular sieve. The absorbed gas is then removed from the molecular sieve by heating.

8. UTILITIES

In most chemical plants the major utilities are steam, cooling water electric power, and air. In addition it is not uncommon to find refrigerants such as brine, process water, special water such as deionized or distilled water, hot oil for heating, various gases such as nitrogen or natural gas, and so on. The most important of these will be covered in this chapter.

HEATING SYSTEMS

Steam

The generation of steam for a chemical plant is a complex subject. The boiler can be fired with coal, oil, or gas and the boiler itself may have many different arrangements of tubes, fire boxes, and so forth. Even the auxiliaries (boiler water feed pumps, water-treating systems, etc.) can vary substantially. For this reason the typical system covered here may bear little resemblance to a steam-generating system that might be used. The basic problems however, are somewhat similar and the need to have some understanding of the problems is important.

A simplified layout of a boiler installation is shown in Figure 8 : 1.

The nature of water treatment for use in a steam boiler depends upon the quality of the water available. If it contains carbon dioxide, the reaction will be on the acid side. By passing the water through a degasifier, the carbon dioxide could be reduced from 50--100 ppm (parts per million) to less than 10 ppm by air stripping the carbon dioxide.

The water is then put through a cationic softener such as a sodium zeolite, by which hardness-forming calcium and magnesium salts are converted to sodium salts.

This water would then be taken through an anion exchanger where silica is removed and chlorides and carbon dioxide are converted to

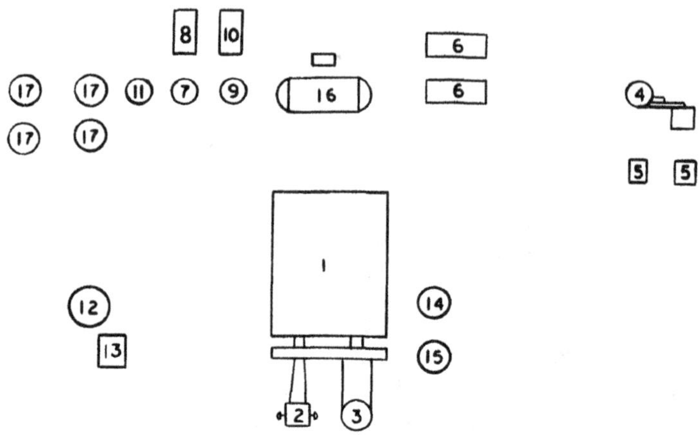

Fig. 8:1. Layout of a typical boiler installation. (1) Boiler; (2) Fan; (3) Stack; (4) Degasifier; (5) Raw Water Pumps; (6) Boiler Feed Pumps; (7) Phosphate Tank; (8) Phosphate Pump; (9) Sulfite Tank; (10) Sulphite Pump; (11) Brine Concentrator; (12) Caustic Tank; (13) Caustic Pump; (14) Continuous Blowdown Tank; (15) Intermittent Blowdown Tank; (16) Deaerator; (17) Water Treatment.

hydroxides. The water is then warmed and combined with condensate return. The hot water is then sprayed into a deaerator vessel so that the air can escape. The condensate return is a very important item. Most of the steam used in a chemical plant is used in heat exchangers and the water resulting from the condensing steam is of good quality. If it is returned to the boiler feed-water system it will reduce the quantity of raw water that must be put through the degasifier, cationic softener, and anion exchangers. A leak in a heat exchanger could, however, contaminate the condensate, so that samples of the boiler feed water must frequently be analyzed for purity.

In addition, sodium sulfite is fed in with the water to react with any oxygen left; and phosphate is added to react with any calcium left and form a salt that will stay in solution.

All of this process, if properly performed, will permit the boiler to produce steam without forming scale in the boiler tubes, and provide soluble, non-corrosive salts, which accumulate as the water is vaporized. Generally, a valve connected to a timer opens for a short time at regular intervals to remove some of the water containing the soluble salts from the drum of the boiler. This is called "blowing down." The higher the loss of condensate, steam leaks, etc., the more water must be added to the system, the faster the salts will build, and the more blow down will be required.

The zeolite softener is regenerated with brine and the anion exchanger is regenerated with caustic.

Hot Oil and Other Heaters

Where a process requires a source of heat over 350°F (176°C), it is frequently impractical to use steam because of the high pressure of steam that would be required. Heating by a natural gas flame would be economical, but it is not always possible or practical, either because of the number of places at which the heat is required or because of the necessity of close temperature control. In these conditions, a hot-oil heater or organic heating system may be used. In the hot-oil heater, a pump can be used to send the oil through a natural gas-fired heater, as shown in Figure 8 : 2.

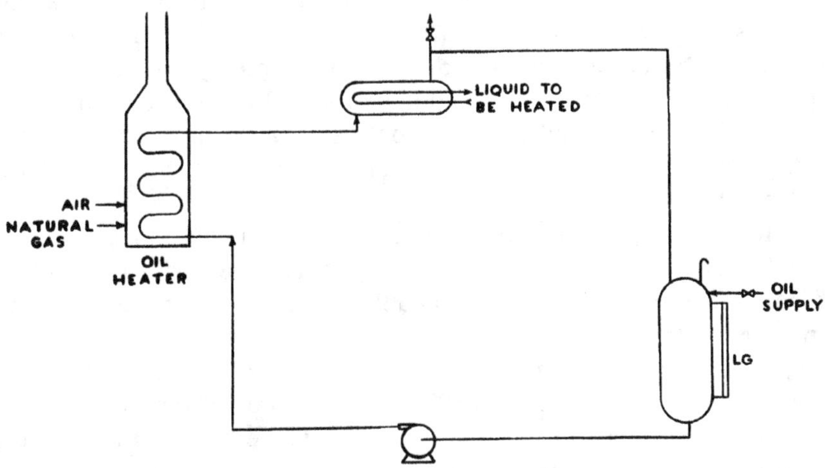

Fig. 8:2. Simplified hot-oil heating system.

To simplify the drawings, the controls on the furnace and vessels are not shown. A surge tank with a vent is provided. The oil expands greatly upon heating and the surge tank must take up this expansion. Any water or air in the oil will produce serious problems and thus the vent at the surge tank is provided for continuous removal of any air or water picked up. Vents must be provided at any high point to allow air to escape initially and in case any of the units is opened for repair.

Another system uses an organic liquid such as Dowtherm (Dow Chemical Company) that will vaporize when heated by natural gas, for example. The liquid is selected that has a low pressure for a given temperature, high flash point, and high latent heat of vaporization. The design of this system is generally similar to that of a steam boiler system.

COOLING SYSTEMS

Cooling Tower

The average chemical plant uses very large quantities of water; therefore it is not economical to use fresh water for cooling and to discard the water. The water could be cooled by recycling it through heat exchangers similar to those that cool the water in an automobile, or by spraying the water in the air from ponds. The average chemical plant uses cooling towers for the recycled water.

In the cooling tower, the warm water returning from the heat exchangers is allowed to cascade over slats. Air moving through these slats either by natural circulation or drawn through by a large fan picks up some of the water vapor. About 1000 BTU of heat are removed for each pound of water that is evaporated. If one percent of the water is evaporated the remaining water will be about 10°F cooler than when it entered.

If the cooling tower depends upon natural circulation or wind it is called an *atmospheric tower.* Where a relatively steady wind of at least three miles an hour is available and the wind is relatively dry (an area of low relative humidity), the atmospheric tower is quite adequate. If a fan is used to ensure good movement of air, the tower is called a *mechanical draft tower.* In some of the older towers the fan pushed the air into the tower. These forced-draft towers are not as efficient as the induced-draft towers, by which air is pulled through the tower.

The induced-draft cooling tower has a cylinder on top to direct the moist air up and away from the tower so that it does not recycle through it. A concrete basin or a reservoir at the bottom of the tower gives surge capacity to the system.

Operating Suggestions

Before starting the fan after a shut-down for repairs, make sure the ball weight is in place in the vibration safety switch.

Inspect the decks or slats carefully for any sign of impurities that would indicate leaks in the heat exchangers.

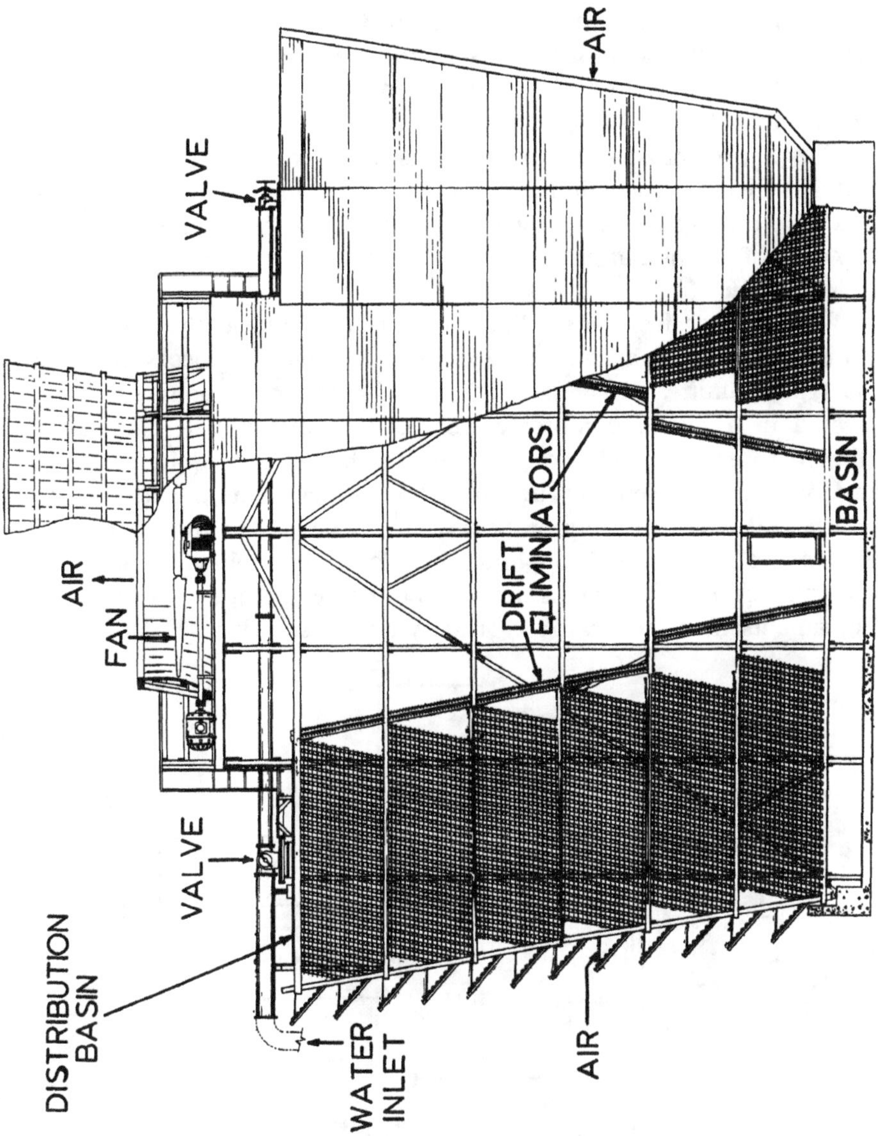

Fig. 8:3. Cutaway of an induced draft cooling tower (*Marley*).

Listen or watch for gas coming from the distributing pipes in the tower, which would indicate gas leaks in a heat exchanger.

Check to make sure the suction screens leading to the cooling water pumps are clean and free of trash.

The testing of samples of water is very important. Unless the water is treated and maintained within certain limits of concentration and pH, fouling of heat exchangers will follow rapidly. Watch for any sign of accumulating algae.

Refrigeration

Where water from a cooling tower is not sufficiently cold for a plant process, mechanical refrigeration is generally necessary. In the typical unit, a refrigerant (such as Freon or ammonia) is compressed in a piston compressor, cooled to condense it, and then allowed to expand and take up heat. Although the liquid Freon or ammonia may be sent through pipe lines to the units and the vapor returned, it is not uncommon to use the refrigerant to cool brine and then pump the brine throughout the plant for cooling. The brine system does not furnish as low a cooling temperature as the refrigerant itself, however. A typical system is shown in Figure 8 : 4.

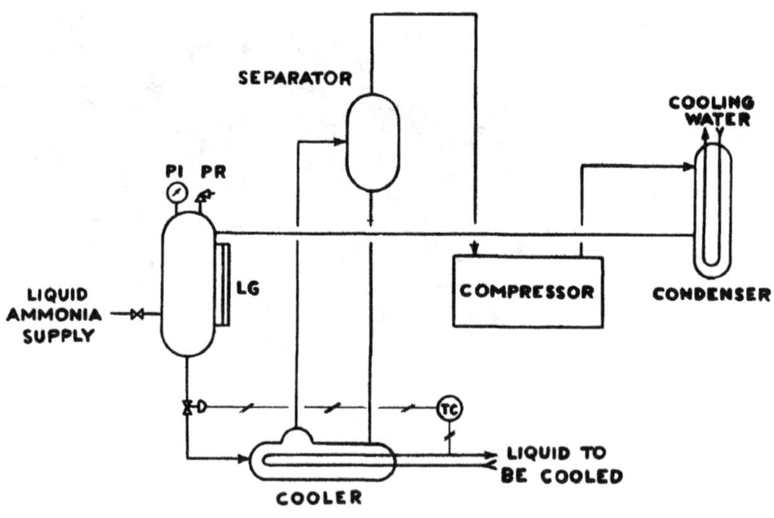

Fig. 8:4. Simplified flowsheet of an ammonia cooling system.

PROCESS WATER

Besides cooling water, a plant will have pure water (for drinking, safety showers, eye-wash fountains, etc.) and process water. The process water may be further classified depending upon the intended use. Where city or well water is used it generally contains dissolved and suspended metallic salts, rust, etc., and it is possible that these compounds can interfere with the purity of the chemicals being manufactured.

The most simple treatment consists of sand filters that remove all but the dissolved salts. Frequently a chemical can be added in very small amounts to react with certain salts and produce a solid that can be filtered out.

When certain ions—from dissolved salts—must be removed, the water can be run through columns filled with ion-exchange resin. If the positive ions such as calcium, magnesium, and lead must be removed, a cationic resin is used; this resin can be regenerated with an acid—for example, nitric or hydrochloric. If the negative ions such as chlorine, fluorine, etc., must be removed, an anionic resin is used that can be regenerated with a base such as sodium hydroxide.

So-called *soft water* is generally run through a cationic resin. If the water goes through both an anionic and a cationic resin bed it is generally called *deionized water* (D. I. water). More expensive but not uncommon is *distilled water.* Distilled water in general costs two or three times as much as D. I. water, depending upon the cost of steam and other variables.

ELECTRICAL SYSTEMS

Even if one has little to do with the design, installation, maintenance, or repair of the electric system, some understanding of the problems is necessary. Generally, we are interested in *horsepower* (a pump motor, agitator, etc., is referred to in terms of horsepower) and one horsepower equals 0.7457 kw (kilowatt) or 745.7 watts. As the watts are equal to the voltage times the amperage, we see that the higher the voltage, the lower the amperage required for a given amount of watts or horsepower.

Most chemical plants have motors for use on 220- or 440-volt circuits to keep the amperage low. This will allow the use of wire of smaller diameter and smaller, less expensive circuit breakers or magnetic starters.

Usually, power is brought into the plant and reduced by one or more transformers to 440—sometimes 220—volts. It then is fed to a panel or box containing a breaker or disconnecting switch. From here it goes to a motor-control center, which is a grouping of magnetic starters. The disconnecting switch and motor control center are frequently located in a separate room, isolated from the chemical processing equipment. The reasons for this is that the electrical equipment, if isolated, need not be in waterproof or explosion-proof cases.

As it would be inconvenient to walk to this electrical equipment room to start or stop a motor, magnetic starters are used. The magnetic starter is simply a large spring-loaded breaker or disconnecting switch held closed by an electromagnet. As the magnet uses very low amperage, the stop—start push-button station (located near to the motor and wires from the push-button station to the magnetic starter) can be small.

In many chemical plants, the stop-start push-button station is in an explosion-proof box (often called NEMA 7 or 9), a waterproof box (often called NEMA 4), or a dust-tight box (often called NEMA 12), and is equipped with a lockout. The lockout generally consists of a small metal bar that can be inserted between two retaining lugs to hold down the stop button. This prevents the accidental starting of a piece of equipment while someone is working on it. The bar should never be removed unless a check shows that everyone is clear of the equipment, and that the repairs are complete so that damage will not result.

Fuses or overload heaters are housed with the magnetic starter. These are intended to protect the motor from burning out, and the current-carrying capacity of each is stated on the motor nameplate. Fuses must be replaced if overloaded. Generally, overload heaters are used rather than fuses, and can be reconnected when sufficiently cool by pushing the reset button. If these protective devices continue to disconnect the equipment, something is wrong and needs correcting. If it is a pump, for instance, it is generally the packing of the pump shaft that is too tight or possibly the discharge of the pump that needs to be throttled back (see section on *Centrifugal Pumps* for explanation). In an agitator also, the trouble might be due to too-tight packing in the shaft.

AIR SYSTEMS

The most general use of air in a chemical plant is to furnish the needs of instruments controlling the process. In addition, it is often used in a stripper, in the process itself to drive air-operated vibrators on bins or hoppers, pneumatic tools, etc. Generally, the air is compressed by a piston-type compressor to about 100 *psig* and cooled to below room or ambient temperature. If the plant is not inside a heated building, additional water must be removed by use of silica gel or similar material so that water will not condense in the lines. The 100 *psig* pressure allows the use of smaller lines to carry the air to the points where it is needed, and it is not uncommon to find that the pressure has dropped to about 80 psig at the point of use because of losses along the line. It is importanece to remove water to prevent freezing in the air lines and to make sure that moisture will not enter and gradually corrode the instruments.

A typical air system is shown in Section 3, which deals with the handling of gases.

The drawing of the air system is simplified. Generally, the compressor has controls (to unload the valves when the pressure in the surge tank is sufficient), cooling water, high-temperature cutoff, etc. The cooler usually has controls for the cooling water and is followed by a knockout chamber with automatic drain. The surge tank is generally equipped with a valve for relief of pressure, and a pressure gauge.

9. INSTRUMENTS FOR PRESSURE

GENERAL

The subject of instruments and controls is rather complex because of the wide selection and combination of instruments available. For this reason we will cover the subject in groups, beginning with the primary elemets for the measurement of pressure. Then will follow temperature, flow rates, liquid level, recorders and controllers, and control valves.

The primary element is an instrument that has some contact with, and reacts to, the temperature, pressure, flow rate, or liquid level. The thermometer and pressure gauge are good examples of primary elements.

Transmitters detect the reaction of the primary element and transmit this reaction in a usable form to a valve, recorder, or controller. The signals sent by the transmitters may be electric or pneumatic. In the pneumatic systems, air compressors send air into a *header*, which distributes the air at a pressure between 50 and 200 *psig* throughout the plant to the instruments. The air is generally reduced in pressure to around 20 *psig* by small pressure regulators located near the instruments. This air is used to power the instruments and to send signals through copper or plastic tubing to other instruments or air-operated valves. The pneumatic instruments have the advantage of minimizing electrical wiring and possible sparking in areas where sparks might cause fires and explosions. The electric instruments have advantages over the pneumatic in some cases. The various advantages and disadvantages of pneumatic, electric, and other forms will be covered in more detail as we consider the individual control systems.

INDICATORS

The device selected to measure pressure is chosen on the basis of the pressure to be measured and whether we are interested in simply knowing the pressure or in having the device send a signal,—either an electrical or an air-pressure signal—to a controller. The following pages will cover the various pressure-measuring devices in order of increasing pressure.

Manometers

A very simple form of pressure-indicator is the *manometer*, which is a glass tube in the form of a U. If this tube has one end connected to the process point to be measured and the other end open to the air, the water or mercury in the tube will move in relation to the difference in pressure. The typical manometer shown in Figure 9 : 1 is filled with water or mercury until both levels of liquid are at the zero mark. If the process point being measured were greater than atmospheric pressure, the level in the left arm would go down and the level in the right arm would go up. If the pressure at the process point were less than atmospheric, the opposite would happen. If the scale is in inches and the levels are one-half inch away from the zero, the total

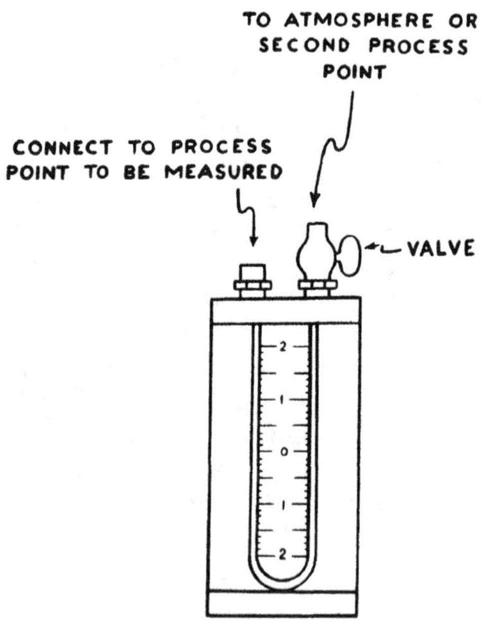

Fig. 9:1. A typical manometer.

difference is one inch. If the manometer is filled with water, this would be 0.0361 pounds per square inch (*psi*), and if it were filled with mercury the one inch would represent 0.491 *psi*. This device is very accurate for measuring pressures from zero to about 30 *psia*.

A manometer can also be connected to two process points to

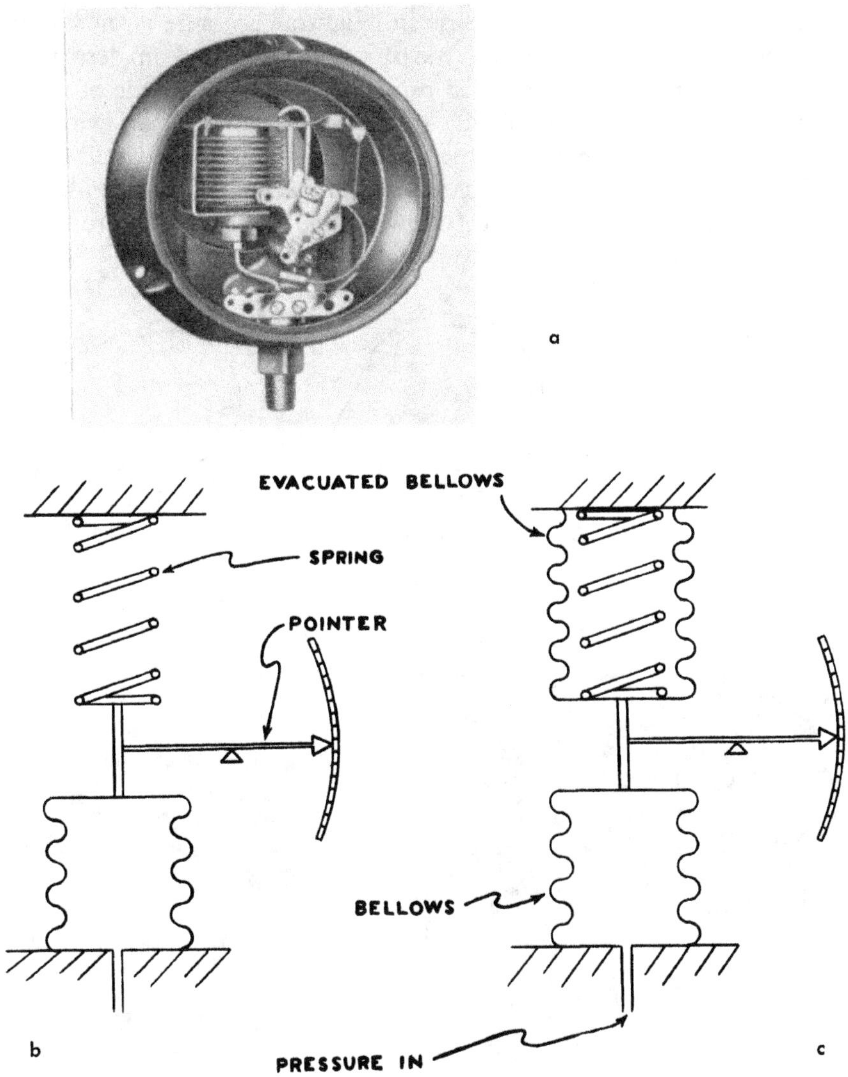

Fig. 9:2. (a) Single bellows in case with face removed; (b) operating principle of a single bellows gauge; (c) double bellows gauge.

measure the difference between them—for example, the drop in pressure across a piece of equipment. The device is limited, however, to a small range of pressures and is generally not adaptable to automatic control.

Bellows Pressure Gauges

A bellows that is sealed at one end and open to the pressure being measured at the other will increase in height as pressure is increased. The bellows gauge is good for measuring pressures from less than atmospheric (0.4 *psia*) to about 25 *psig*. The bellows is made of stainless steel, brass, or monel metal. It may be attached to an arm that will move a pointer, pen, or other device to generate a signal; thus, the bellows type gauge can be used as a simple indicator or with recorders or controllers. A single bellows gauge shown in Figure 9 : 2a

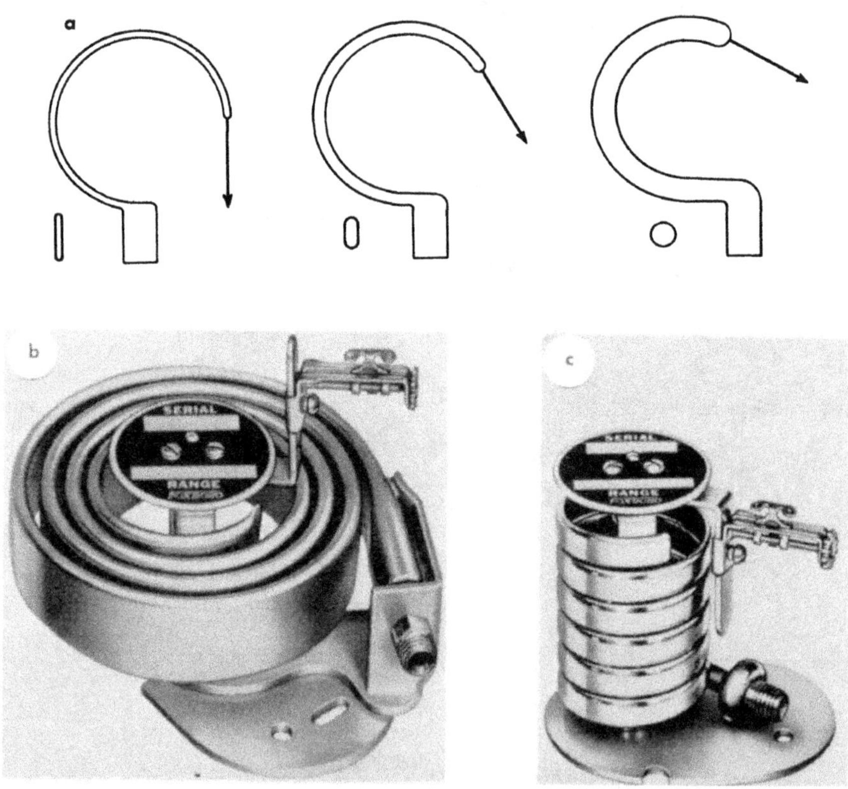

Fig. 9:3. (a) Reaction of a bourdon tube to pressure; (b) spiral and helix type (*Foxboro*).

measures pressures above atmospheric. If the spring were enclosed in a bellows from which the air had been evacuated, the pressure measured would be absolute pressure (*psia*). The double diaphragm gauge shown in Figure **3** : 2b is generally used for measuring pressures between 2 *psia* and 25 *psia*.

Bourdon Tubes: Spiral and Helix

Pressure gauges using formed tubing that tends to straighten out as pressure is applied are in widespread use. The most common (and inexpensive) is the Bourdon tube. Figure 9 : 3 shows the operation.

The *spiral* is actually a series of Bourdon tubes and thus has greater movement, which eliminates the need for mechanical amplification of the movement. The *helix* is also a series of Bourdon tubes having substantial movement for a given change in pressure. These last two types are shown in Figure 9 : 3.

Bourdon tubes can be made from steel, bronze, stainless steel, and various alloys. These gauges have wide ranges—from zero to about 100,000 *psia*; however, they are not as accurate as the manometer or double bellows for pressures below 15 *psig*. The gauges also have greater accuracy in the middle third of their range; thus, 5 *psig* on a 0-100 *psig* gauge is not too accurate; reading 5 on a 0-15 *psig* gauge would be much more accurate. Unless a gauge is periodically tested, the reading can vary considerably from the true pressure. Generally,

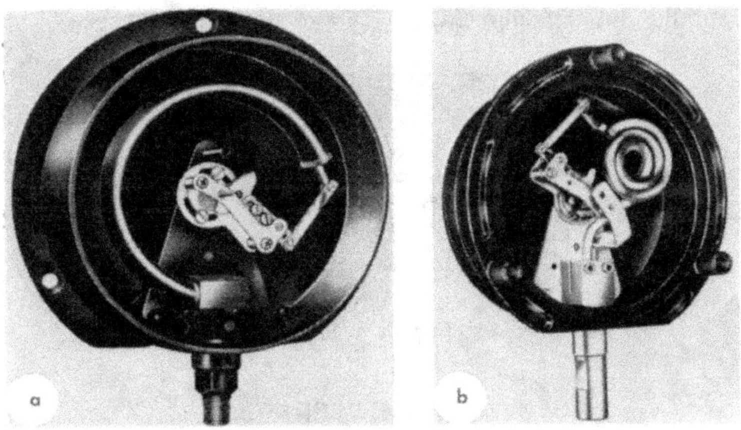

Fig. 9:4. Pressure gauges with faces removed. (a) Bourdon tube; (b) spiral type (*Ashcroft; Manning, Maxwell & Moore*).

a valve is placed between the pressure gauge and point to be measured to allow for removal for testing and resetting.

Gauges that indicate pressures both below and above a zero mark—the zero is atmospheric pressure—are called *compound gauges*. The reading above zero is generally pounds per square inch gauge (*psig*), whereas the reading below zero is generally inches of mercury. A reading of 10 below zero is therefore 10 inches, or about 5 *psi* vacuum or 10 *psia*.

Gauges are specified by the diameter in inches of the face or scale, and the range of pressure to be measured; the common sizes are 2, 3½, 4½, 6, 8½, and 12 inches.

A Bourdon tube and a spiral for simple indication of pressure are shown in Figure 9 : 4.

Diaphragm Seal

Often it is necessary to protect the gauge from the material that is being measured. With steam, a loop of tubing is placed between the gauge and the steam point. This loop of tubing gathers water and prevents the steam from entering the gauge. With chemicals that might corrode the gauge, chemicals that might cool or freeze, or very viscous chemicals that might plug the tube, a so-called *chemical gauge* may be used. This is the usual gauge except that a *diaphragm seal* (Figure 9 : 5) is placed between the gauge and the chemical, and the space between the diaphragm and gauge is filled with oil or similar material to transmit the pressure. The necessary liquid is added at the factory. The gauge should not be taken apart and the plug for filling should not be removed.

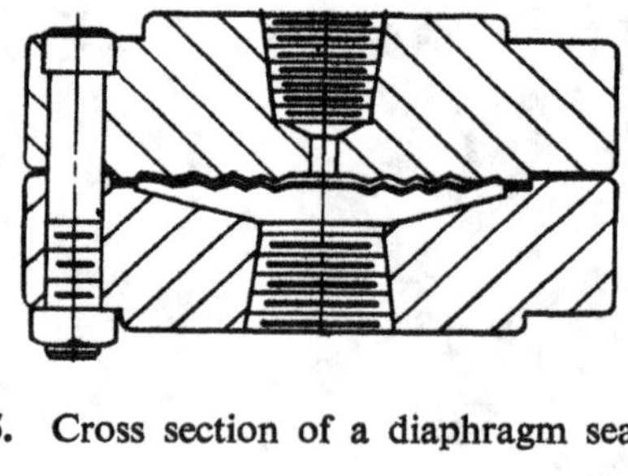

Fig. 9:5. Cross section of a diaphragm seal (*Foxboro*).

TRANSMITTERS

Where we wish to record the pressure or control the process from a distant location, a pressure transmitter can be used. These are of

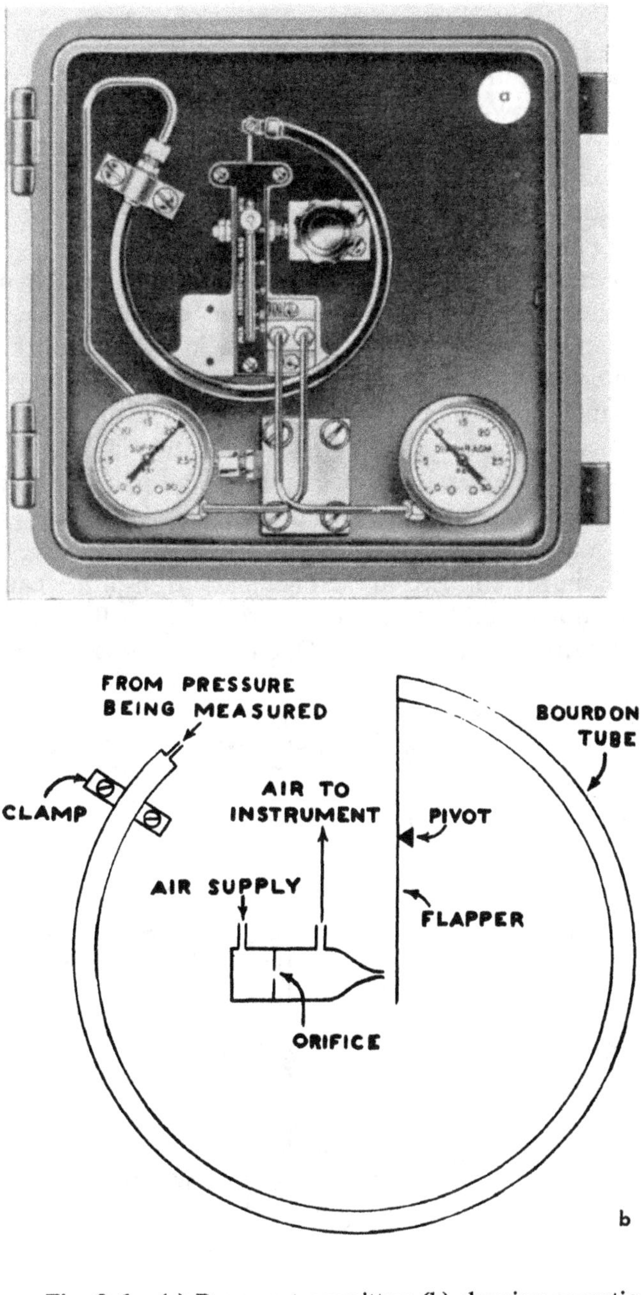

Fig. 9:6. (a) Pressure transmitter; (b) showing operation. (*Fisher-Governor*).

several types:

Pneumatic Transmitters

In a pneumatic transmitter, the reaction of the primary element is changed into a low-pressure air signal that can be used by a recorder or controller. Most recorders and controllers need an air signal of 3 to 15 *psig*. The transmitter can also be equipped with a pointer so that in addition to sending a signal to the recorder or controller an indication of the pressure is given.

These pressure transmitters may vary according to the manufacturer, but they generally have supply air at about 20 *psig*, which is limited in flow by an orifice or restriction in the line. As pressure increases, a bellows, Bourdon tube, or spiral or helical primary element moves a flapper or flat piece of metal in front of small opening through which the supply air is escaping. As the opening is blocked, the pressure builds up in the copper tubing and is transmitted to the recorder or controller. A typical arrangement is shown in Figure 9 : 6.

Electrical Transmitters

The same movement used in the pneumatic transmitter can be used to move the loop of a differential transformer that will convert mechanical energy into electrical energy. One name for this item is a *Dyna-*

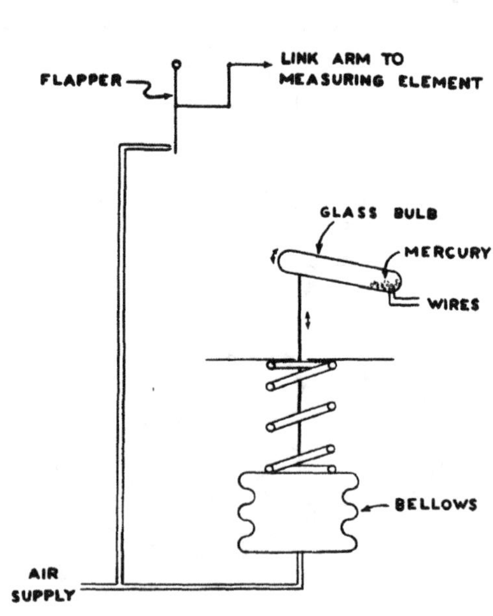

Fig. 9:7. Diagram of a mercury pressure switch system.

former transducer.

Another item, widely used, is the *mercury pressure switch.* Essentially, this consists of two wires running into a sealed bulb that contains mercury at one end. The bulb can be tilted and if tilted so that the mercury comes down to the end where the two wires are located the electrical circuit is completed. These mercury pressure switches can be used to set off an alarm if pressure becomes too high or too low. They can also operate a solenoid valve or the magnetic starter of a large piece of equipment such as a compressor.

In the mercury pressure switch (shown in Figure 9 : 7), the mercury bulb making contact is held in this position by the air pressure on the bellows. If the link arm, moved by a Bourdon tube or similar device, moves the flapper away from the nozzle, the air pressure in the bellows will drop, the end of the bulb will come down, letting the mercury run away from the two electric wires, and the electrical circuit will be broken.

Electronic Transmitters

An electronic system is fast and accurate, particularly for pressures that must be recorded from a thousand feet to twenty or thirty miles away. With this system, 115-volt, 60-cycle power is supplied to the transmitter. A Bourdon tube or similar device exerts a force on a balance, which calls for additional electric power to put the balance back in the level position. An eletronic tube picks up this power and rectifies it to direct current; this direct current, varying between one-half and five milliamperes, can be carried by relatively small wires to a recorder or controller many miles away.

Figure 9 : 8 illustrates the system. The balance beam (2) is held horizontal when the torsion (pull) exerted by the Bourdon tube (1) is equal to the pull of the magnet (3). The magnet derives its pull from the amount of electric current being put out as a signal. The signal must therefore balance the action of the Bourdon tube. If these do not balance, the beam of the balance will move. Any movement by the beam toward or away from the oscillator coil (4) will cause the inductance of the coil to change. This in turn will cause an increase or decrease in the current that is being picked up and rectified by the electron tube (5), and thus change the signal being emitted.

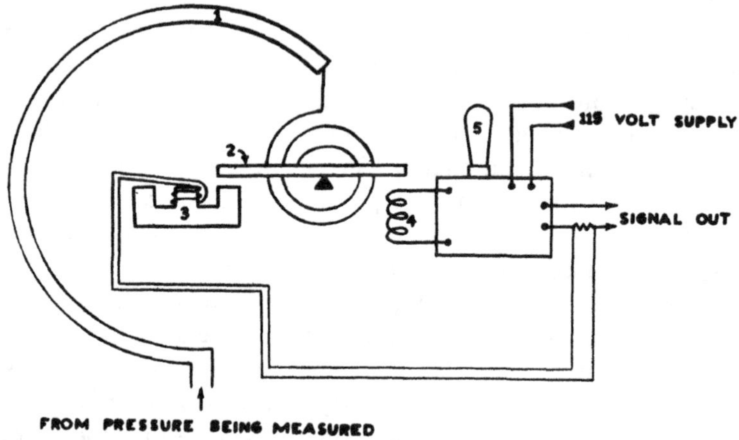

Fig. 9:8. Diagram of an electronic transmitter.

REGULATORS

A pressure regulator can sometimes combine the functions of detector, transmitter, controller, and control valve.

Frequently a high-pressure supply of gas must be furnished in reduced pressure. An example would be a supply of natural gas to a furnace or boiler; or the very common air supply for instruments, which is stored and transported by pipe line throughout the plant at pressures of 50-100 *psig*. This pressure must be reduced to 20 *psig* for use by the instrument. Pressure regulators are purchased for a limited range of supply pressure and are adjustable for a limited range of output pressure. They are also purchased for the volume of gas they can handle. A typical instrument for regulating air supply is shown in Figure 9 : 9

The air supply is attached (7), flows through the screen (11) and then through the valve (4). Air from the outlet side of the regulator places a pressure on the diaphragm (8). When this pressure is sufficient to overcome the pressure of the spring (9), the diaphragm flexes upward and permits the small spring (5) to push the valve upward to stop the air flow. If the outlet pressure goes down, the diaphragm flexes down and forces the plug out of the seat to permit high-pressure gas to flow. Screwing the knob (1) down increases tension on the spring, thus requiring higher pressure on the output side in order to

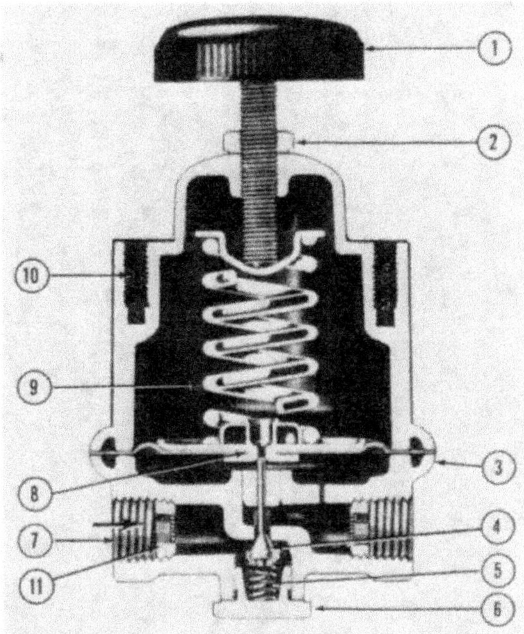

Fig. 9:9. Cutaway of a pressure regulator (*Foxboro*).

close the valve. A lock nut (2) prevents excessive pressure. The diaphragm is held in place by being clamped between the body halves (3). The regulator can be mounted in a control panel using screws into the body (10). Cleaning can easily be accomplished by removing the bottom plug (6).

10. INSTRUMENTS FOR TEMPERATURE

PRIMARY ELEMENTS

Expansion Type—Thermometers

The mercury thermometer is still probably one of the most economical and simple devices for measuring temperature. In the chemical industry it is used in the form shown in Figure 10 : 1a. Not only mercury but also other liquids, gases, and vapors will expand when heated. Although these other substances cannot be seen, like the mercury in a thermometer, they exert a pressure when they expand, and if this pressure is measured by a gauge, we can determine the temperature. The range of visual indication of the expansion of mercury is generally from −35 to 1000°F. The range of the gas- or liquid-filled temperature indicators, where pressure is measured, is −300 to 1,400°F, depending on the material used. These latter can be directly connected, or the measurement bulb and gauge can be connected with tubing up to 150 feet in length. Several types are shown in Figure 10 : 1.

Thermocouples

A *thermocouple* is a device consisting of two wires of different metals connected together that will generate electricity when the common end is heated. As this electric potential depends upon the difference in temperature between the measurring (joined) and the reference (open) ends, the potential can be measured in millivolts (one millivolt equals 0.001 volts), and from this the temperature can be calculated.

The followng is a list of different thermocouples in common use:

Positive Wire	*Negative Wire*	*Maximum Temp. °F*	*Millivolts/ °F*
Platinum-rhodium	Platinum	3,000	0.0075
Chromel	Alumel	2,220	0.0215
Iron	Constantan	1,400	0.0290
Copper	Constantan	600	0.0295

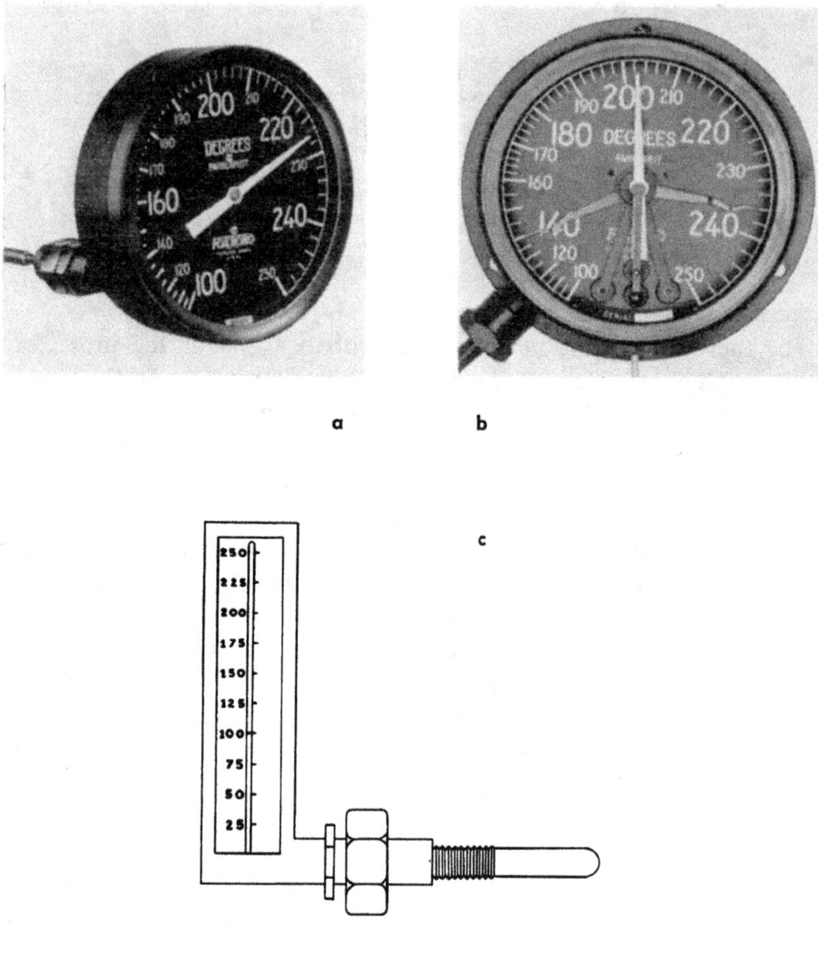

Fig. 10:1. Temperature indicators. (a) Gas bulb type (*Foxboro*); (b) gas bulb type with high and low alarm contacts (*Foxboro*); (c) mercury type.

The millivolts of electric potential developed for one degree Fahrenheit are average for the maximum temperature shown, but the important point is that an instrument calibrated to record temperature by measuring millivolts from thermocouples of copper and constantan will not give the correct result if a thermocouple of iron and constantan is connected to it. The minimum temperature measurement for thermocouples is about 0°F, except for copper-constantan, which will go down to —300°F with reasonable accuracy.

Because of corrosion, the thermocouple is generally enclosed in a well to protect it from the chemicals, and a head provided for running the wire from the thermocouple to a recorder. A typical unit is shown in Figure 10 : 2.

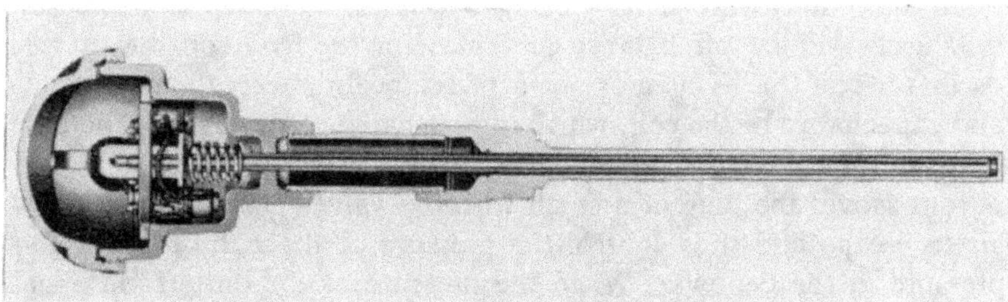

Fig. 10:2. Cutaway of a thermowell (*Foxboro*).

Resistance Elements

The electrical resistance of a wire will change with the temperature of the wire, and this can be used to measure temperature. A piece of glass, mica, or plastic is wound with a length of nickel or platinum wire and is generally placed in a well to protect the wire from corrosion. This type of device is generally more accurate than the thermocouple, but the range of temperature is smaller, and it must be carefully protected from any corrosion. The range of this type is generally limited to —40 to 600°F.

Radiation Elements

The radiating type of element can be used for temperatures of 800° to 3200°F. The radiant energy or heat from a furnace or other source being measured is focused by a mirror or lens onto a series of thermocouples called a *thermopile*. The output of the thermocouples can then be measured, and the temperature determined.

TRANSMITTERS

Where the primary element consists of a bulb filled with liquid, gas, or vapor and exerts a pressure on a Bourdon tube, the same types of transmitter used with pressure will serve. For thermocouples, the potential generated by them could be run to a recorder, but there are times when it might be desirable to convert the electric potential of the thermocouple to a pneumatic signal.

Pneumatic Transmitters

An electro-pneumatic transmitter is shown in Figure 10 : 3. The blanace beam (1) is held horizontal by the current from the thermocouple (2), causing the magnet (3) to pull down on one end of the beam. The output air signal (from the transmitter) causes a bellows (4) to expand and exert a balancing pressure on the balance beam by a torsion spring.

If the temperature that is being measured increases or decreases, the beam will be off balance and move away from or toward the oscillator coil (5). This movement of the beam affects the inductance and capacitance of the coil, which in turn causes a decrease or increase in the output of rectified direct current from the electron tube (6). This output moves the plug of a small solenoid valve (7), which will allow air to escape and thus control the pressure of the output signal and pressure in the bellows. When the pressure of the output air signal is sufficient to balance the action of the thermocouple, the beam is back in balance, and everything will remain constant until a change in temperature occurs.

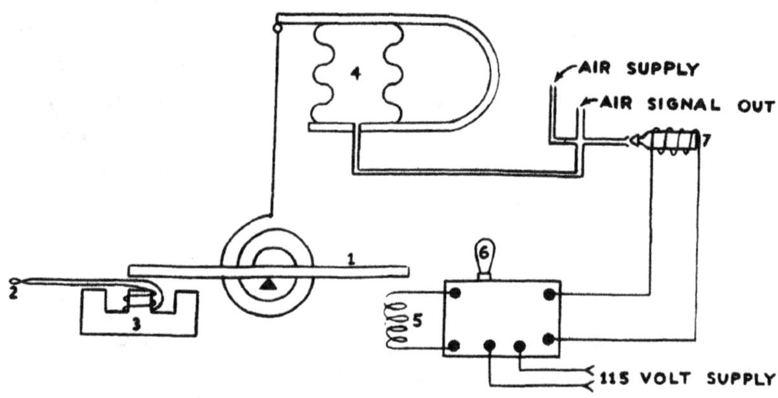

Fig. 10:3. Diagram of an electro-pneumatic transmitter.

Electronic Transmitters

Although thermocouples may be used and the potential generated by them run to recorders, there is a limit to the distance. For long distances, the potential should be converted to a proportional direct current, which is independent of the resistance of long lengths of wire.

Figure 10 : 4 shows an electronic arrangement using a balance beam. The thermocouple (2) attracts the beam (1) while a portion of the output signal (4) is used to repel the beam at the magnet (3). If a change in temperature occurs, the beam will be off balance, and the end of it will move toward or away from the oscillator coil (5), thus changing the inductance and capacitance of the coil, which will change the rectified direct current being put out by the electron tube (6).

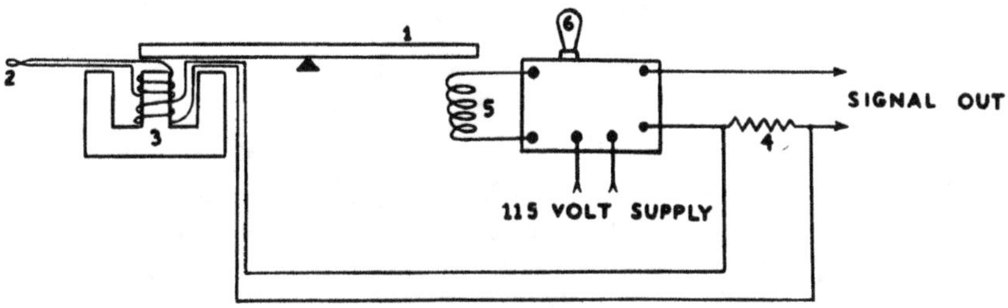

Fig. 10:4. Diagram of an electronic transmitter.

REGULATORS

A combination regulator can be purchased that will control the temperature by controlling the flow of steam or water. Like the pressure regulators, they have a reasonably narrow range. They are generally available for standard sizes of lines from 0.5 to 1.5 inches. They are puchased for the required range of temperature, e.g., 10-70°F, 50-120°F, 80-150°F, 110-170°F, 130-190°F, 140-200°F, and 200-250°F.

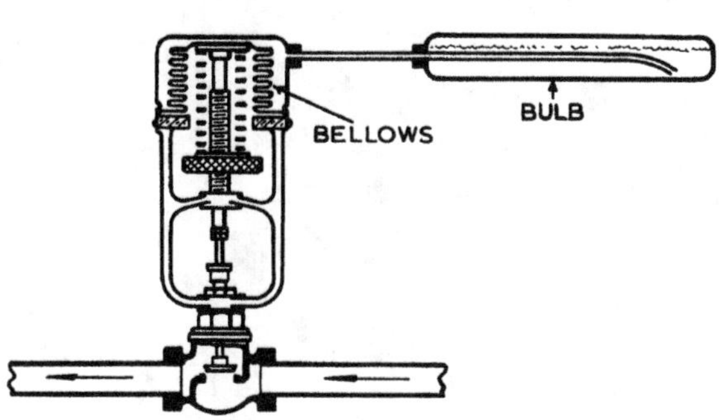

Fig. 10:5. Cutaway of a self-contained temperature regulating valve (*Fulton Sylphon Division, Robertshaw Controls*).

Generally, a mercury thermometer is used with these units to set and check them. For cooling water, they are designed to open on increasing temperature; for steam services, on decreasing temperature. A typical valve is shown in Figure 10 : 5.

The bulb is inserted in the vessel to be controlled. The liquid or vapor in the bulb exerts a pressure on the bellows, which in turn opens or closes the valve.

11. INSTRUMENTS FOR FLOW AND LIQUID LEVEL

FLOW INDICATORS

Bull's Eye

The simplest type of flow indicator is the *Bull's Eye* (Figure 11:1). It does not give the exact rate of flow, but it will indicate whether the material is flowing and give a rough idea of the rate of flow.

The bull's eye can be obtained in the form of a check valve, if desired. It is generally purchased for the size of pipe with which it is to be used and can be flanged or screwed to the pipe. Temperatures to 400°F and pressures to 500 *psig* are satisfactory for sizes under one inch . With the six-inch size, the temperature and pressure are limited to 100°F and 100 *psig*.

Fig. 11:1. Screwed flapper (check valve) bull's eye (*Fischer & Porter*).

Pitot Tube

The pitot tube is a piece of tubing that is inserted into a pipe and turned to receive the full impact of gas moving in the pipe. A static tube is similar, but the end is closed and a small slot cut in the side of the tube. This tube, combined with the pitot tube, is generally referred to simply as a pitot tube. The static tube measures only pressure while the pitot tube is effected by velocity of the gas; thus the difference between the two will indicate the velocity of the gas. Such a device is satisfactory in many gas pipes or clean fluids, but in general it must be calibrated in use by some other means and is not too accurate. A complete pitot assembly is shown in Figure 11 : 2.

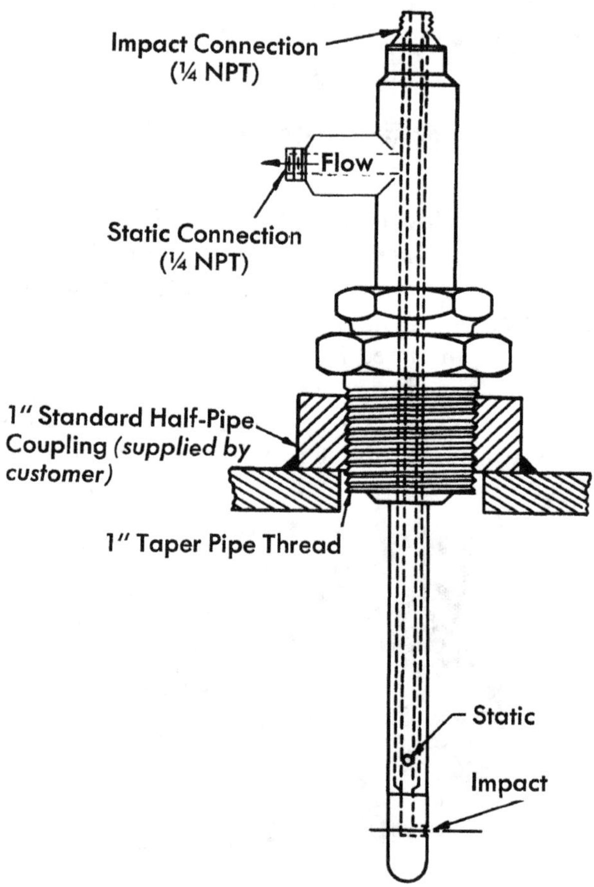

Fig. 11:2. Cutaway of a pitot assembly (*Foxboro*).

Venturi

A venturi is a piece of pipe in which the diameter has been decreased for a short distance in order to increase the speed of a gas, which will decrease its pressure. As applied to an instrument in the chemical industry the term is limited to a special decrease in the diameter of the pipe or tubing. The tube is tapered in at an angle of about 25 degrees then out at an angle of about 7 degrees; the pressure drop created by this gradual restriction will then indicate the rate of flow. Venturis are difficult to manufacture and are generally limited to applications where viscosity is high, where there is considerable suspended solids, or maximum accuracy is required (Figure 11 : 3).

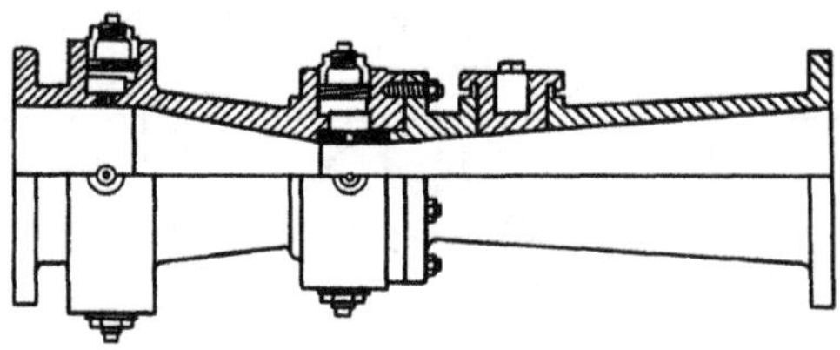

Fig. 11:3. Cutaway of a venturi (*Foxboro*).

Dall Tube

The Dall tube (Figure 11 : 4) is similar to the venturi except that it is shorter and less expensive. Use of it is generally limited to clean fluids where low pressure drop across the venturi is desired.

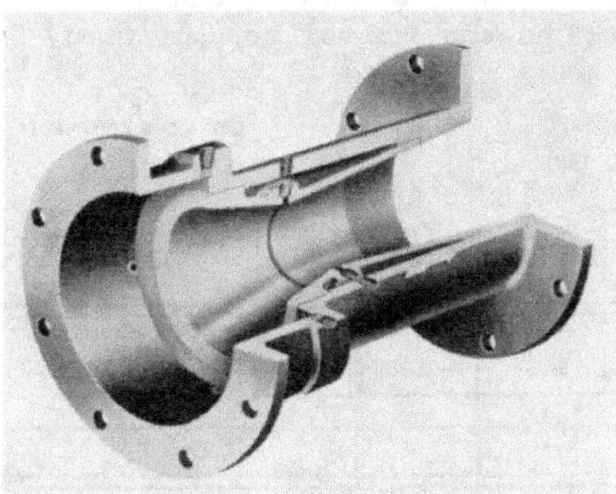

Fig. 11:4. Cutaway of a Dall tube (*Foxboro*).

Nozzle

In place of the venturi, a *nozzle* (Figure 11 : 5) can be used to produce the same effect and is much less expensive to manufacture. In this device also, drop in pressure is measured to find the rate of flow. The device will handle very small quantities of solids in the stream. The greatest advantage is a low pressure drop for a given flow.

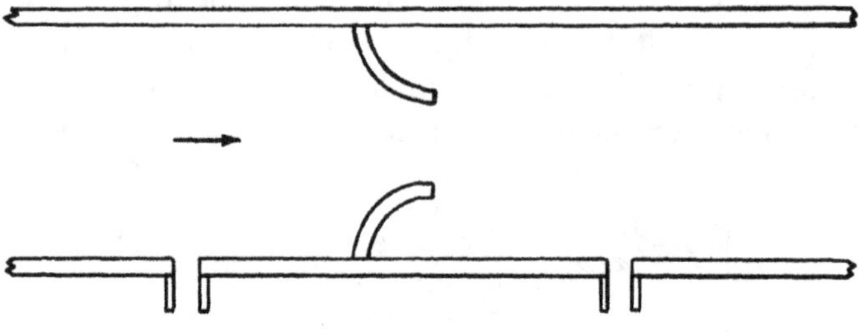

Fig. 11:5. Cutaway of a nozzle.

Orifice

An orifice is essentially a flat plate with a hole very accurately cut into it. The restriction to flow causes a pressure drop that will be related to the rate of flow in the pipe. As a given orifice can be used for only a relatively small range of flow, flanges are used to hold the plate in position and allow orifices with different size holes to be inserted easily.

The disadvantage of an orifice is that it cannot be used with solids and that it has a higher pressure drop than the units previously mentioned. It is however the most popular primary flow element

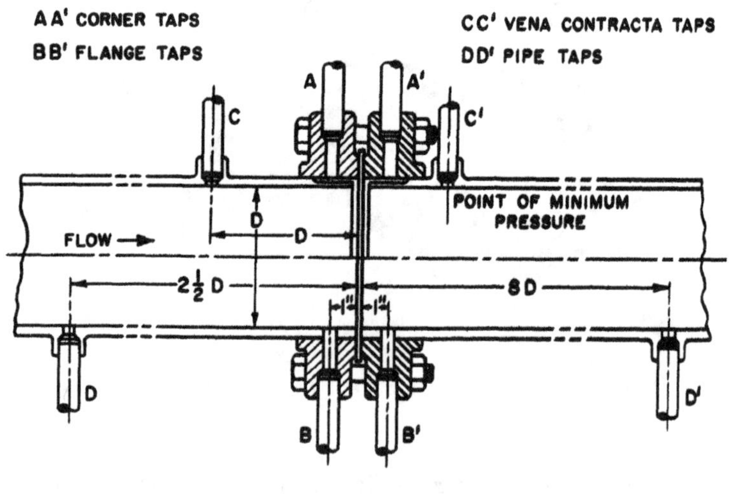

Fig. 11:6. Various connection systems for an orifice (*Foxboro*).

because it can be manufactured cheaply and generally the higher pressure drop required is not a serious problem.

Tap points, that is, threaded holes or couplings from which tubing runs to the instrument that measures the pressure drop across the orifice, should be located with proper consideration given to the size of pipe, pressure drop expected, and range of the instrument.

In Figure 11 : 6, flanges at A and A′ where pressure is measured are called *corner taps*; B and B′ are called *flange taps*. In place of these, the pressure could be measured by tapping into the pipe at point C′ (distance to the orifice, half the diameter of the pipe), and C (distance from the orifice, the same as the diameter of the pipe); these are called *radius*, or *vena contracta*, taps. The pipe could also be tapped at point D′ (distance to the orifice, 8 diameters) and D (distance to the orifice, 2.5 diameters); these are called *pipe taps*.

With a given rate of flow, the difference in pressure between the flange taps (BB′) is greater than between the radius taps (CC′), and the difference in pressure of the radius taps (CC′) is greater than that of the pipe taps (DD′). For lower rates of flow, flange taps would be used; and for high rates, radius or pipe taps would be used, so that the range of differential pressures (DP) expected is correct for the range of the DP cell (see page 123).

Corner tap flanges are sometimes used for small pipes (two inches or less), and in this case the taps are almost next to the plate of the orifice, rather than being about an inch away, as are standard flange taps (Figure 11 : 6).

Area Meter: Rotometer

For low liquid flows—from about 0.05 to 200 gpm—the area meter, or *rotometer*, is an inexpensive instrument to use. Essentially this consists of a tube—the tube is glass for pressures not exceeding 450 *psig* in smaller sizes—that increases in cross-sectional area, and a ball or float of known weight. As flow increases, the ball or float rises in the tube. A scale, generally etched on the tube, will indicate the rate of flow of material.

For higher pressures, the rotometer may be armored or enclosed in a steel case with a window, or made completely of steel with a magnetic coupling to indicate rate of flow, as shown in Figure 11 : 7. As the float rises, the armature rises. An electromagnet is placed around the armature and sends an electric signal in porportion to the height of the armature.

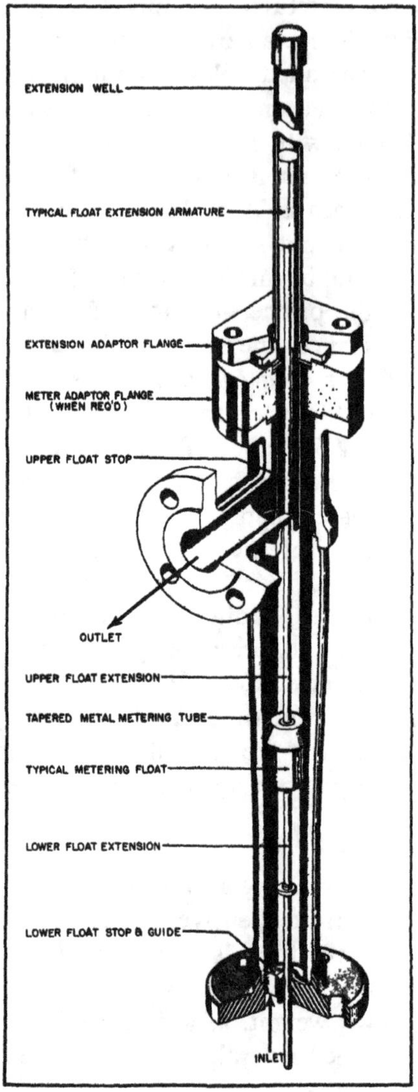

Fig. 11:7. Cutaway of a steel area meter (*Fisher and Porter*).

FLOW TRANSMITTERS

Pneumatic

A number of devices or primary elements can be used in a line to indicate the rate of flow in terms of a difference in pressures. A manometer could be attached to the two pressure points and with a proper scale the rate of flow could be read. A form of manometer that uses a float can serve to drive an arm with a pen and thus record the flow locally.

DP Cells

Two pressure transmitters could be used to send signals representing each of the pressures, but special *differential pressure* (*DP*) *cells* are available that simply transmit the difference between the two pressures.

Up to 32 *psi* difference in pressure, and static pressures to 6,000 *psig*, the two pressures can be applied to both sides of a diaphragm, which will flex toward the side of lower pressure. Attached to the diaphragm is a force bar that controls the distance between a nozzle and a flapper. This, in turn, controls the pressure that is built up and transmitted; see Figures 11 : 8 and 11 : 9a.

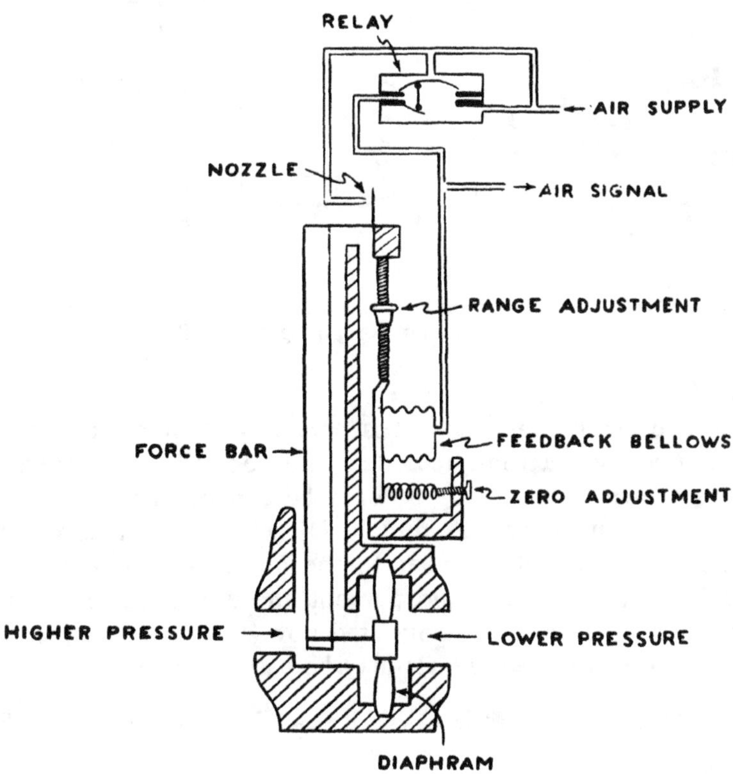

Fig. 11:8. Cutaway of a DP cell.

Figure 11 : 9b is a diagram of piping for a DP cell. The valves on the vertical lines permit the instrument to be removed for repair while the plant is in operation. The valve between the two vertical lines is an equalizing line for setting the instrument. With no flow, or with the valve between the vertical lines open, the DP cell should be adjusted so that no signal is emitted. This setting or adjusting is called "zeroing" the instrument.

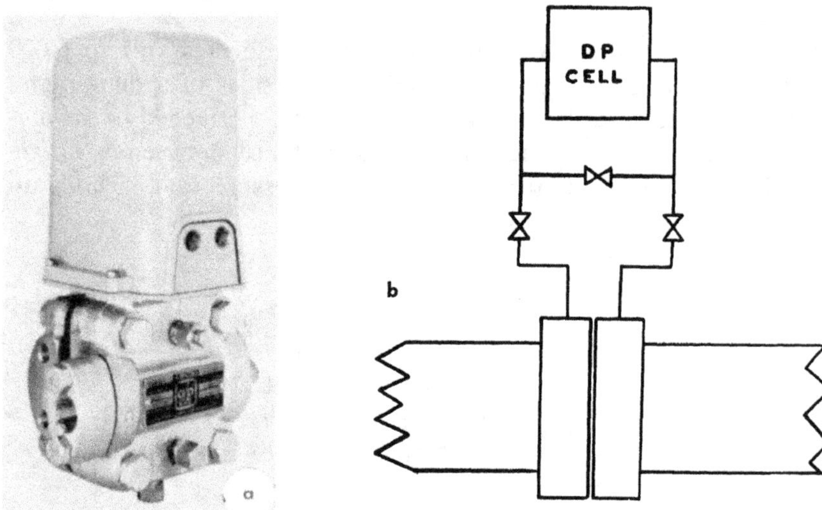

Fig. 11:9. Differential pressure cells. (a) standard model (*Foxboro*); (b) diagram for piping of a DP cell.

Electronic

There are situations where it is necessary to obtain a signal that can be used by electronic recorders and controllers. A transmitter such as shown in Figure 9 : 8 could be connected to a DP cell to convert the pneumatic signal from the DP cell to a direct current signal. An inductive rotometer similar to the one shown in Figure 10 : 7 could be obtained that would put out an electric signal. Although these devices are not commonly used now, their use may increase in the future because their signals may be used by computers.

LIQUID LEVEL INDICATORS AND TRANSMITTERS

Glass

The simplest device for indicating the level of a liquid is the glass tube connected to the bottom and top of a vessel. The level in the tube will be the same as the level in the tank. Combination metal and glass level gauges are available, called *reflex* gauges, that can withstand high pressures, are safe, and are easy to read. The problem is that solids, if present, can plug these gauges and cannot transmit a signal that can be used to control the level.

Bubbler

For tanks open to the atmosphere, a simple bubbler (Figure 11 : 10) is quite effective for measurement and to furnish a signal to control the level. A low flow of air or gas is allowed to bubble up through the liquid from a tube that extends nearly to the bottom of the tank. As the level of liquid builds up in the tank, the pressure of the gas in the top of the tube will increase. This pressure can either show the level or be transmitted to an instrument that controls the level.

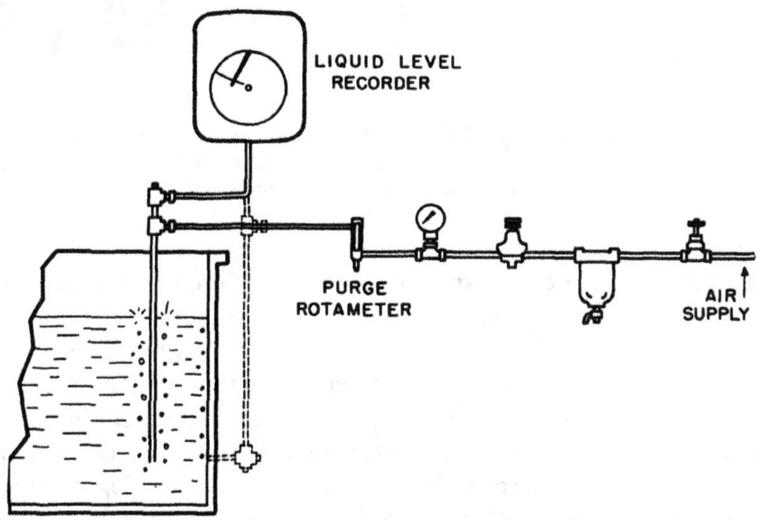

Fig. 11:10. Diagram of a bubbler system with recorder.

Floats

Floats in tanks open to the atmosphere are simple and cheap. Tanks under pressure require a more complex float, as shown in Figure 11 : 11a. These floats are limited to chemicals that will not corrode the float and sealing bellows. Solids that would collect in the float chamber or build up on the float itself will upset the accuracy of this device. The float can be purchased as a float-transmitter system, as shown in Figures 11 : 11a and b

Fig. 11:11. (a) Cutaway of float compartment of a float-transmitter; (b) face of transmitter with cover removed (*Mason-Neilan Division-Worthington*).

DP Cells

The measuring and transmitting of the difference between two pressures to indicate the level of a liquid is very similar to the problem of measuring flow and, indeed, the same DP cell may be used.

A special DP cell has been developed that will simplify the piping because it can be mounted directly on the tank. Depending upon the circumstances, the leg going to the top of the tank may or may not be filled with liquid, but generally it is safer to have this leg filled and avoid the problem of accidentally getting liquid into the leg. If the leg is not filled and liquid gets over into it, the pressure shown on the leg will go up and thus indicate a lower level in the tank than it should.

This DP cell is operated in exactly the same way as the DP cell used for measurement of flow and shown in Figures 11 : 8 and 11 : 9b. The piping is shown in Figure 11 : 12. If the lower pressure-connecting line on the right is filled with liquid, a net range depression will be required rather than a range elevation. Valves are not shown on the piping drawing.

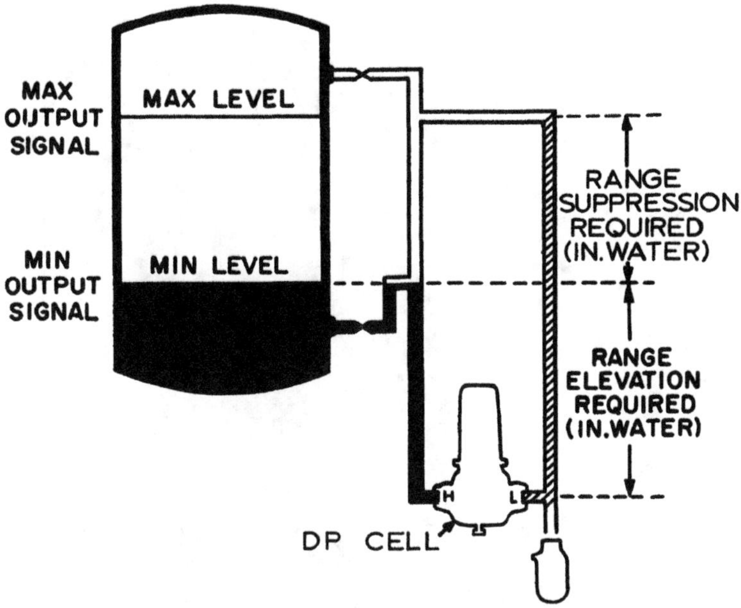

Fig. 11:12. Piping diagram for using a DP cell as a level transmitter (*Foxboro*).

12. INSTRUMENTS FOR CONTROL

RECORDERS AND INTEGRATORS

As has been explaned, a pneumatic or electrical signal can be received from a transmitter that will be proportional to temperature, pressure, flow, or level. These signals can now be used to make a record on a chart. For flow, this signal can also be used to drive an integrator or totalizer that will keep a running total of gallons of liquid or cubic feet of gas.

Two major types of chart are used—*circular* charts or *strip* charts. The charts may be driven by a spring, which must be wound each day like a watch, they might use a synchronous electric motor as in the electric clock, or they may be operated by pneumatic drive. As these recorders are in many ways simplified versions of the recorder-controller, they will be discussed later in this chapter.

CONTROLLERS

To understand a controller, it is first necessary to define a few terms:

Set point is the value to be maintained, such as a temperature, pressure, level, or rate of flow.

Control Point is the value the controller is able to maintain, which may or may not be the same value as the set point.

Offset is the difference between the set point and control point when conditions are stable.

Deviation is the difference between the set point and control point when conditions are unstable.

Load is the job or work that must be accomplished in the process, such as heating, cooling, or pumping.

Reaction Time is the time required for the process to adjust to a new load (increased or decreased load) after the controller has changed (opened or closed) the control valve.

Transport Lag Time (or distance-velocity) is the total time required for the process to adjust from a change in load to the new load.

Controllers may be equipped with devices to provide smoother, more accurate, or faster control. The need for these devices and method by which they achieve their objectives is not easy explain, but a reasonable understanding of the subject is necessary for anyone connected with a modern chemical plant. These various devices are discussed in the following text.

Proportional Action

The simplest form of controller is the *on-off controller*. Any change in load will result in the controller opening or closing a valve completely. Adding proportional action (sometimes called proportional band) to a controller will give smoother control by requiring less change in the control valve with a given change in the load. The following example will show the results that proportional action can achieve.

Figure 12 : 1 shows a simple problem of controlling streams of hot and cold water coming together. A control valve is in the hot water line and the temperature of the combined streams will be held at 60°F by controlling the air pressure to the control valve; that is, controlling the degree to which the valve is open.

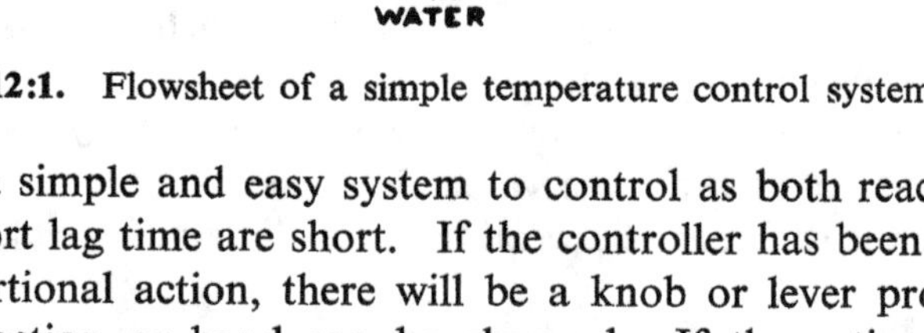

Fig. 12:1. Flowsheet of a simple temperature control system.

This is a simple and easy system to control as both reaction time and transport lag time are short. If the controller has been equipped with proportional action, there will be a knob or lever provided by which the action or band can be changed. If the action is set at minimum or zero, the controller will be an on-off controller, as though it had no proportional action. As the action or band is increased, the controller will change the pressure to the control valve less for each degree of temperature by which the warm water differs from the desired temperature.

For example, if the desired temperature were 60°F and the proportional action were set at 50%, the control valve might be fully closed when the temperature reached 85°F and fully open when the temperature dropped to 35°F. If the proportional action had been set at 10%, the valve might be fully closed when the temperature dropped to 55°F. Unless the flow of water were being stopped and started, the temperature would not cycle to the extremes where the valve was fully opened or closed. Under normal conditions, where the use of warm water was gradually increased or decreased the valve would only open or close a little more.

Considerable offset can occur and the reason for this can be explained by further consideration of the above example. If, with a 50% proportional-action setting, the controller were holding the valve one-third open to obtain 60°F warm water, and flow of cold water were increased, the following would occur.

1. The primary element would send a lower pressure signals to the controller.
2. The controller would send a higher pressure air signal to the control valve to open it further.
3. The flow of hot water would increase.
4. The temperature of the warm water would increase.

The temperature of the warm water would not reach the desired 60°F because the valve must now be more than one-third open. Since at 60°F the primary element put out a signal to the controller, which then sent an air pressure signal to the valve to hold it one-third open, it cannot hold the valve more than one-third open at 60°F. The system must therefore stabilize at a lower temperature. This new temperature would be about 54°F, depending upon the amount of increase (load change) in the flow of cold water. The valve would be almost half open.

If the proportional action had been set at 10%, the system might have stabilized at 57°F rather than 54°F. At 57°F the primary element would have sent less of a change in signal, but the lower proportional action would have produced a greater reaction of the controller, which would have opened the valve further. It may therefore be seen that the greater the proportional action, the greater the offset will be.

If the proportional setting had been 2% instead of 50% or 10%, the temperature might have stabilized at 58°F but the controller would have been reacting very quickly to the slightest change and would have produced cycling. This cycling is frequently called "hunting" because the valve moves up and down as if hunting for a point of control.

If all the preceding had been recorded on a graph, the picture would looke like Figure 12 : 2 for the different settings. Line A is the 50% setting, which caused excessive offset. Line B is the proper setting for this example. Line C is the too-narrow setting of 2% that caused cycling.

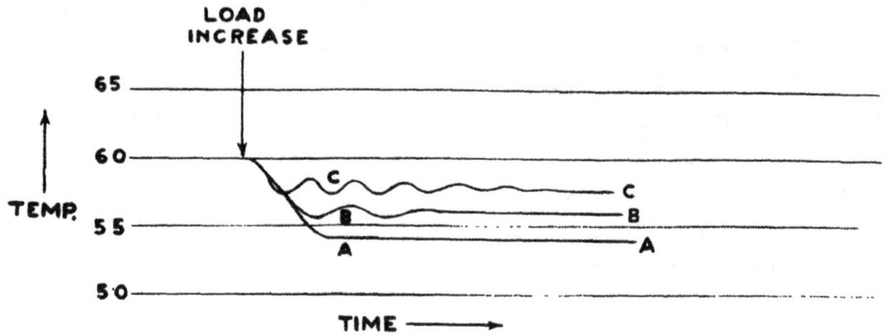

Fig. 12:2. Graph showing the effect of proportional action setting on offset and cycling.

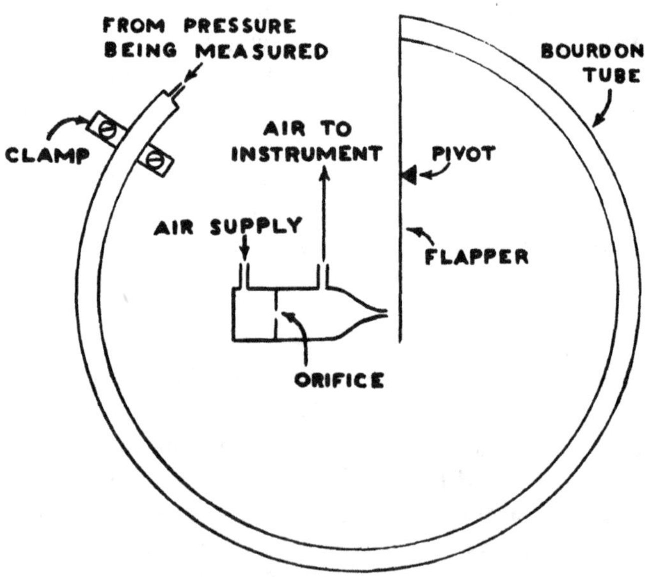

Fig. 12:3. Diagram of a pressure-actuated controller with proportional action.

A controller with proportional action is shown in Figure 12 : 3. A Bourdon tube receives a pressure signal from the process. This could be from a temperature bulb, DP cell, or similar primary element or transmitter. Supply air at a pressure of about 20 *psig* is furnished to the instrument as a source of power. An orifice limits the quantity of air so that the device does not consume an excessive quantity of air. As the pressure from the process increases, the Bourdon tube will begin to straighten and the metal flapper will begin to move closer to the small air opening of the air chamber that allows the supply air to escape. The air pressure will then increase in the air chamber and a higher pressure of air will be sent to the control valve. A knob can move the fulcrum so that the flapper moves more or less for a given movement of the Bourdon tube. The movement of this fulcrum provides the proportional action setting.

The pressure-actuated (pneumatic) controller is easily reversed by changing the air chamber to the other side of the flapper so that it puts out a higher air signal as pressure decreases. This is called *reverse acting*. The controller is purchased for maximum ranges of 50, 100, 250, 500, 1,000, etc., *psig*, and should be so selected that the figure desired is about the middle of the range.

Pressure-actuated controllers are also available, in which a bellows-and-spring is used in place of the Bourdon tube for pressures below 25 *psig*. Where the controller also indicates pressure by a pointer or records it by means of a pen, the spiral is often used in place of the Bourdon tube, because the movement of the spiral is greater for a given change in pressure.

A pressure controller that also indicates pressure is generally called a *pressure-indicating controller*, or PIC. One that records on a chart is referred to as a pressure-recording controller, or PRC. Both circular-and strip—chart recorder controllers are available and are shown in Figure 12 : 4.

Reset Action

For many control problems in the modern chemical plant, the change in load may be gradual, but reaction time and transport lag are usually considerable. In such cases, adding reset action to the controller can help proportional action to control the process at the set point. The mechanical method used to obtain reset action will be considered later in this chapter, but the nature of reset action and what it accomplishes can be studied by the following example.

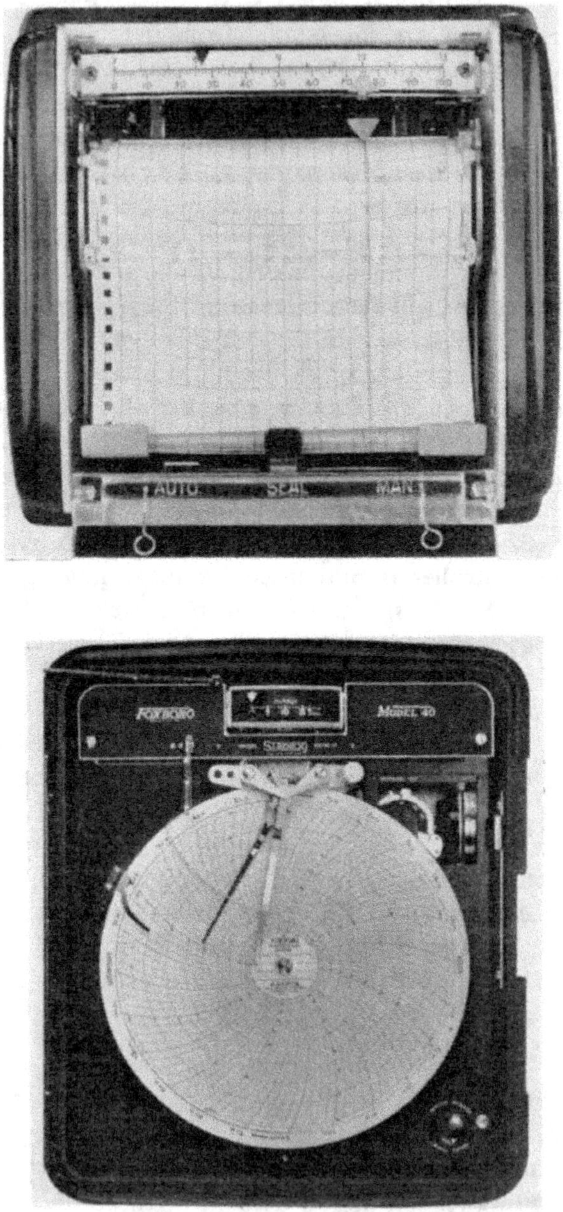

a

b

Fig. 12:4. (a) M-54 recording controller for strip chart; (b) M-40 recording controller for circular chart (*Foxboro*).

As shown in Figure 12 : 5, liquid is being heated in the tubes of a heat exchanger by steam. If the steam-control valve opens, the heat must be transferred from the steam to the tubes, through the wall of the tubes, and into the body of the liquid being heated. The reaction time here is much greater than in the first example studied, and if proportional action only were used, a narrow setting would produce severe cycling, while a wide setting would produce considerable offset.

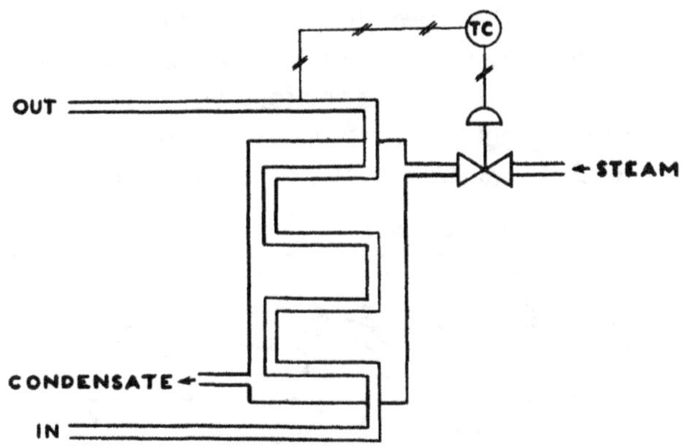

Fig. 12:5. Flowsheet of a typical temperature control system.

To control the temperature at 200°F the controller might keep the control valve one-third open. The heat load is now increased by either a greater flow of liquid to be heated or by the entering liquid being colder than before. To correct this the control valve would need to be half open. As in the previous example where there was no reset action, the temperature would stabilize at a lower point and the control valve would not be one-half open. Here is where the reset action starts.

Reset time may vary from about one third of a minute to 150 minutes. For example, if reset time is one minute, one minute after the temperature goes below 200°F the reset action would increase the pressure to the control valve. If the system had begun to stabilize at 195°F, the reset action would not be satisfied because the system was below the control point desired; it would add a small amount of output pressure to the control valve and repeat the addition each minute. When the temperature reached 200°F the reset would stop adding pressure, and if the temperature goes above 200°F it would subtract pressure from the output. When stability is finally reached with reset action, although the controller is receiving the same signal as a result of the 200°F temperature, it would be putting out a greater signal to the control valve than previously.

As with proportional action, too much or too little reset action can cause difficulties. In addition, since the reset action reinforces the proportional action, they must be set in relation to each other. Figure 12 : 6 shows the lines that would appear on a chart if the temperatures were being recorded in the example cited. Line A shows the effect of too-slow a reset action, which results in long periods of time to recover or reach a stable temperature. Line B is about right; and line C shows cycling as a result of too fast a reset action.

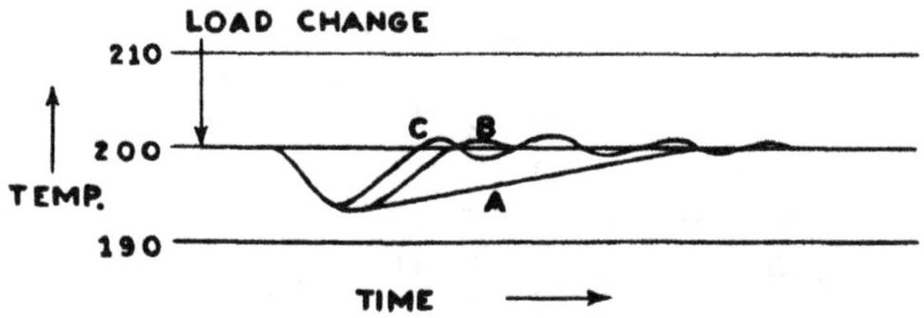

Fig. 12:6. Graph showing the effect of reset action setting on recovery time and cycling.

Rate Action

The third function that can be added to a controller is rate action, frequently called derivative control action. Rate action is needed in order to obtain a substantial râte of correction to counteract substantial changes in load.

A good example is a furnace used to heat oil, which in turn is being used to heat processes in the plant. If the reboiler of a large fractionating column should suddenly be cut in, it could make a serious change in the heat load required. As the furnace itself holds a good deal of heat, the initial load change would barely be detected by the primary element. The gas valve would open very little and the furnace and oil temperature would continue to fall. It would require considerable time until the gas valve were open sufficiently to add the necessary heat. This could upset temperatures throughout the plant.

Even with reset action, the recovery time would be long because the time of reaction rate is large, thus requiring a slow setting of the reset action to prevent cycling.

With rate action, the fact that the temperature was going down at a faster rate all the time would cause rate action to take over and open the gas valve fully, even though only a minute or two had passed and the temperature might be only a few degreees from the control point. The graph in Figure 12 : 7 shows the effect of too much rate action (line A), sufficient (line B), and none (line C).

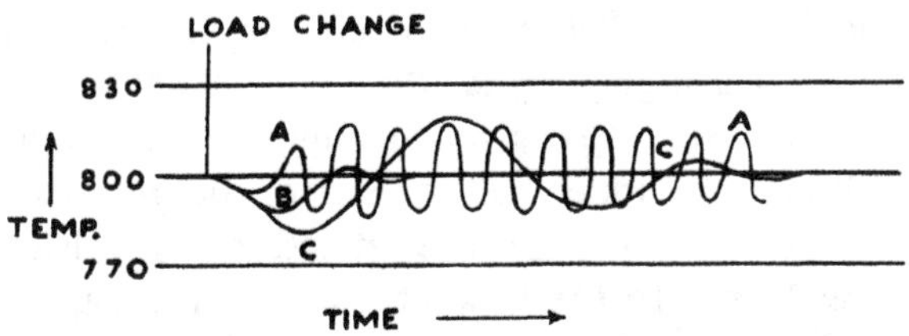

Fig. 12:7. Graph showing the effect of rate action setting on recovery time and cycling.

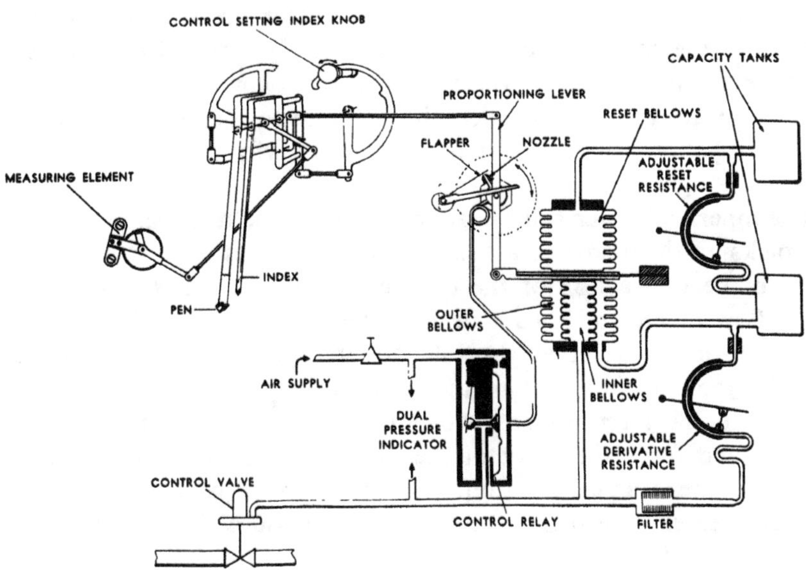

Fig. 12:8. Diagram of Model 40 recorder-controller with proportional, reset, and rate action (*Foxboro*).

Controller, Fully Equipped

There are many different types, combinations, and arrangements of controller. A reasonably popular pneumatic model with proportional band, reset, and rate action is shown in Figure 12 : 8.

When the actuating signal from a transmitter pressure, temperature, level, or flow causes the measuring element to move, the pen and proportioning lever will also move. If the proportioning level moves to the right, the flapper approaches the nozzle and the pressure to the control relay is increased. This change in pressure is small but it moves the diaphragm of the control relay to the left and opens the valve, thus allowing the air supply to increase pressure to the control valve and inner bellows. The control relay is therefore acting as an amplifier to give out a higher signal than it receives.

This increased pressure also goes through the filter and derivative resistance. Although the inner bellows senses the increase in pressure very quickly, the outer bellows senses it less quickly because of the restrictions or resistance to the flow of air. With a rapid increase in pressure, this difference becomes greater and air flows into the outer bellows. With the help of the outer bellows (increased cross sectional area pushing up) the support and pivot of the proportional band lever is moved upward. This moves the flapper and striker arm toward a more vertical position and produces a greater pressure in the diaphragm of the control relay. In other words, the rate action reduces the proportional band. If there is no further change in signal from the transmitter to the controller, the pressure to the control valve, to the inner bellows, and to the outer bellows will finally become the same.

The inner and outer bellows are opposed by the reset bellows. This also picks up the increased air signal after it reaches the outer bellows. As with the rate action or the derivative bellows, there is resistance to the flow of air and thus a delay in reaction.

It was mentioned that the rate action reduced the proportional band by moving the pivot of the proportioning lever. In time, the pressure in the reset bellows will equal that of the inner and outer bellows, and force the pivot of the proportioning lever to return to essentially the original position. The pressure in each would, however, be greater than before, and thus the pressure signal to the control valve would be greater.

OPERATING POINTERS

In general, when setting the controller with proportional band, reset, and rate action (derivative), the following procedure should be followed:

1. Set reset and rate (or derivative) action to minimum, by setting the derivative adjustment to minimum time and the reset adjustment to maximum time.
2. Adjust the proportional band by reducing it from maximum value until cycling occurs. Remember to make these adjustment slowly and allow the controller and the process time to reach equilibrium.
3. Increase rate action (derivative) in steps until cycling disappears and then reduce it until cycling stops.
4. Reduce proportional band further to produce cycling and increase the rate action.
5. Repeat the steps until further increase of derivative time fails to stop the cycling introduced by reducing the proportional band.
6. Increase the proportional band just enough to stop the cycling.
7. Adjust reset at the same value as derivative or slightly less. In any case, use a minimum setting without causing any cycling.

CONTROL VALVES

As mentioned previously, a change in temperature, pressure, flow rate, or liquid level can be detected and transmitted and a resultant corrective air signal be obtained from a controller. Generally, the air signal is sent to a control valve where the air moves a diaphragm that opens or closes the valve. The valves are purchased much as other valves are, with consideration of pipe size, connections such as screwed or flanged construction, materials, pressure, etc. In addition, there are some special considerations for control valves.

Depending on the purpose it is to serve in a certain process, the valve may be purchased to close or to open with increasing air pressure. This is generally a safety consideration in case the air supply should fail. A control valve in a steam supply line might be purchased to close upon failure of air, whereas one on cooling water might be purchased to open on failure of air. In general, a valve with air to open is closed by an air signal of 3 *psig* from the controller, half opened by 9 *psig*, and fully opened by 15 *psig*.

The valves may also be purchased in *single port*, *double port*, *needle*, and other styles. In addition, they can be purchased with packing glands, special sealing glands, packless bellow seals, and sections of heat-dissipating fins (see Section 4). Where the process requires only an on-off controller, a single valve with standard seat control may be used. Where the process requires wide proportional band and close control, a double-seated "V" type of port valve should be used. A cutaway of a typical control valve is shown in Figure 12 : 9.

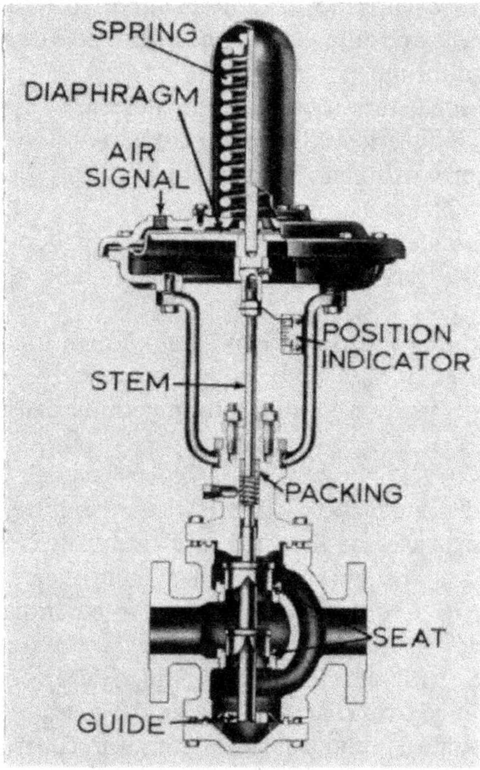

Fig. 12:9. A diaphragm-operated double-port control valve (*Foxboro*).

MISCELLANEOUS

Valve Positioners

On large control valves and control valves on high pressures, or where the control valve has a tendency to stick, a valve positioner can be used. There are various valve positioners on the market, and these can be attached to most control valves. The valve positioner receives the 3- to 15-pound signal from the controller and is also supplied with instrument air pressure (usually between 50 and 100 pounds). The valve positioner uses the high-pressure air to move an arm attached to the stem of the valve, or applies high-pressure air to the diaphragm of the control valve. This moves the valve opening into proper position in relation to the 3- to 15-*psig* air signal being received from the controller.

MANUAL STATIONS

It is sometimes necessary to have a valve that can be adjusted from a remote location. For this purpose a manual air-loading station can be purchased. This unit is much like a pressure regulator. Supplied with air-header pressure, a knob can be turned down to give an output from 0—20 *psig* to the valve. A pressure gauge on the outlet of the manual air-loading station will then show how far open the valve is. If the gauge reads 3 *psig* or less, the valve is closed; at 9 *psig*, the valve is half open; and at 15 *psig*, the valve is usually completely open. On a valve that closes with increasing air pressure (air to close valve) the reverse would be true.

Index

W

Z

www.ingramcontent.com/pod-product-compliance
Lightning Source LLC
LaVergne TN
LVHW011519200926
848484LV00001B/79
9780820600864